Language and Mathematics Education

Multiple Perspectives and Directions for Research

T0344524

A volume in
Research in Mathematics Education
Barbara J. Dougherty, *Series Editor*

Research in Mathematics Education

Barbara J. Dougherty, *Series Editor*

The Classification of Quadrilaterals: A Study in Definition
By Zalman Usiskin

A Decade of Middle School Mathematics Curriculum Implementation: Lessons Learned from the Show-Me Project
Edited by Margaret R. Meyer and Cynthia W. Langrall

Future Curricular Trends in School Algebra and Geometry: Proceedings of a Conference
Edited by Zalman Usiskin, Kathleen Andersen, and Nicole Zotto

The History of the Geometry Curriculum in the United States
By Nathalie Sinclair

The Intended Mathematics Curriculum as Represented in State-Level Curriculum Standards: Consensus or Confusion?
Edited by Barbara Reys, University of Missouri-Columbia

Mathematics Curriculum in Pacific Rim Countries—China, Japan, Korea, and Singapore: Proceedings of a Conference
Edited by Zalman Usiskin and Edwin Willmore

Language and Mathematics Education

Multiple Perspectives and Directions for Research

Edited by

Judit N. Moschkovich
University of California, Santa Cruz

INFORMATION AGE PUBLISHING, INC.
Charlotte, NC • www.infoagepub.com

Library of Congress Cataloging-in-Publication Data

Language and mathematics education : multiple perspectives and directions for research / edited by Judit Moschkovich.
    p. cm. – (Research in mathematics education)
  Includes bibliographical references.
  ISBN 978-1-61735-159-4 (pbk.) – ISBN 978-1-61735-160-0 (hardcover) – ISBN 978-1-61735-161-7 (e-book)
1. Mathematics–Study and teaching. 2. Language arts (Higher) I. Moschkovich, Judit N.
  QA135.6.L37 2010
  510.71–dc22

                                                            2010038047

Printed in the United States of America

CONTENTS

Foreword vii
  David Pimm
Preface xiii
  Judit N. Moschkovich

1 Language(s) and Learning Mathematics: Resources,
   Challenges, and Issues for Research 1
   Judit N. Moschkovich

2 Developing a Mathematical Vision: Mathematics as a
   Discursive and Embodied Practice 29
   Kris D. Gutiérrez, Tesha Sengupta-Irving, and Jack Dieckmann

3 Language in Mathematics Teaching and Learning: A Research
   Review... 73
   Mary J. Schleppegrell

4 Function and Form in Research on Language and
   Mathematics Education 113
   Guillermo Solano-Flores

5 Recommendations for Research on Language and
   Mathematics Education 151
   Judit N. Moschkovich

Afterword.. 171
  Beth Warren

Contributors 175

FOREWORD

David Pimm

What a difference forty years can make. To look at academic research journals in mathematics education now, one could be forgiven for thinking that it has always been the case that augmented verbatim transcripts of classroom or other interactions, that autograph student work in its myriad forms, that teacherly inscriptions in public spaces (such as boards or, latterly, screens) have always been both the focus and manner of reporting in academic study. It can be hard to remember, not least when faced with the enormous amount and range of research pertaining to questions relating to language and mathematics education discussed and framed in this fascinating book, that it was only in the 1960s that even brief teacher–student interchanges (a couple of lines, of uncertain provenance and opaque verisimilitude) or perhaps an example of written student work were offered to readers, whether professional or academic, and only then to augment the account. It seems a far cry from the intense study of a few seconds of a videotaped record, calibrating gesture and speech production (e.g., Sabena, Radford, & Bardini, 2005), or self-styled "gold-standard" audio taped transcriptions and their conversation analyses being read into evidence (e.g., Jefferson, 2004).

One issue has always been that of the language's authenticity and its use in authenticating a truth. When I worked at the Open University in the UK in the 1980s and 1990s, the newly formed Centre for Mathematics Education was involved both in undertaking research studies of mathematics classrooms and in producing course material for practicing teachers.

Language and Mathematics Education, pages vii–xii
Copyright © 2010 by Information Age Publishing

Working with professional BBC sound and film crews (as one of the ultimate products had to be of television broadcast quality), we would tape in each school for a week at a time. This extensive filming (over 90 hours for a four-hour broadcast) was in part a response to a desire for greater diversity and equity of representation, but also reflected teacher complaints when we had shown interviews with one or two students made in a studio setting (the better to be able to hear what they said and to see what they wrote as they wrote it). They said this was all well and good, but what were the other 28 students doing? There is another divide, akin in part to the divide between "everyday" and "formal" mathematics discussed in this book, between representations of classrooms that will be recognized as authentic by teaching professionals and those representations that are useful for academic research. A key element of this involves the nature of the classroom language, about which we all have strong intuitions and expectations.

There is interesting work about this issue being undertaken by a group headed by Sandra Crespo at Michigan State University (Crespo, 2006; Crespo, Oslund, & Parks, 2007) and, independently, by a group at Simon Fraser University in Canada (Zazkis, Liljedahl, & Sinclair, 2009), involving student and novice teachers' ability to "generate" plausible or "realistic" classroom dialogue as part of lesson planning. Another change that has materialized in the last decade is sample exchanges (or even longer scripts) showing up in teachers' editions of textbooks. There are many aberrations I am hearing that these are causing, with, for instance, students being penalized for giving the "wrong" answer (i.e., going off-script—a script that they are likely not even aware exists).

One thing to bear in mind for both of these distinctions (everyday and formal mathematics, authentic representations for teachers and researchers) is that even "realistic" theatre is a genre like any other with its own conventions and aesthetic. "Everyday mathematics" and its practices, whatever these are, are not to be preferred because they are somehow more "natural" or truer; they are simply different (shaped by different intentions and constraints). The same is also the case for the French distinction—due to Chevallard—between "academic" and "school" mathematics (see Love & Pimm, 1996, for a discussion of this point).

In this regard, it is worth attending to the notion of antecedent genres (Jamieson, 1975), where a historic genre shapes a subsequent evolving one (rather like Marshall McLuhan viewing metaphor as inherently conservative and backward-looking, seeing the new in terms of the old). Because we were making records, parts of which would be shown on television, television conventions came into play and framed our sensibilities and awareness, as well of those of the television professionals with whom we were working. How much is watching a classroom tape conditioned by televisual expectations, simply because it is viewed on a TV screen? Likewise, many early transcripts evoke the familiar play script genre (e.g., clean, well-marked turns, non-overlapping speech), so much so that even Lakatos's (1976) *Proofs and Refutations* has been mistaken for a classroom transcript (Pimm, Beisiegel, & Meglis, 2008) rather than the play script whose form I believe it was intended to mimic. And might the same be true with classrooms (wait your turn, don't all speak at once, . . .)?

I raise these issues here to draw brief attention to the question of truth (or veracity or authenticity), when the phenomenon is some aspect of actual classrooms, most significantly here some linguistic or semiotic element of them, which is the core concern of this volume. Even filming in classrooms in the 1980s (with a crew of eight or nine) brought us into contact with some of these issues of conflicting features and values (e.g., there must be no dead air time, so as not to disorient viewers), that we could on occasion shade or fabricate the facts to present a greater Truth (e.g., the reverse time order of certain shots). A fact may be granted the status of fact simply by someone or something (camera, tape recording) attesting that it has occurred. But there are other facts worthy of attention. There is no unmediated truth.

My published work on issues on the interrelationship between language and mathematics has covered nearly thirty years. It is still a most productive place for me to dwell, I find, and this book attests to the current vibrancy of the field, as well as to the complexity it has engendered. More than most, and as mentioned in a number of chapters here, it requires an interdisciplinarity of a particular sort. In a recent paper (Pimm, 2009), I observed:

> Presmeg's paper presumes that what is transferred between disciplines are theories (and 'theoretical frameworks') and methods (and 'methodologies'), not least when in relation to anthropology she emphasizes "particularly with regard to methodology and construction of theory". This has not been my experience. While I am certainly interested in the awarenesses, attentions and practices of a discipline and how they are theorized, what I have transported across various frontiers has usually been individual notions, both identifying and conceptualizing phenomena. (p. 156)

I offer that observation about my own history here as a partial counterpoint. Linguistics itself has been for me a productive place of engagement, one that I feel is underrepresented in the broader mathematics education literature. It can be a challenge getting work drawing on technical linguistic ideas published within mathematics education. Yet a glance at the extensive bibliography of this book shows a number of instances of mathematics education authors placing articles on mathematics education in more mainstream linguistics or linguistics and education publications (e.g., Barwell,

2009). Some of the authors in this book do not see themselves as working in mathematics education *per se*. There is considerable value and benefit for us who work centrally in mathematics education in them coming the other way.

Mathematics, its understanding and expression, does place significant pressures on a language and not only on the technical register but also the wider grammar. The existence of such a register also can perturb wider features of the language (Barton, Fairhall, & Trinick, 1998). But if there is one thing in particular I would value in workers in this area, it would be a greater sensitivity to language itself, its nuance and subtlety, the ways in which it is shaped and can itself shape the reality it both describes and brings into being.

Whenever a term gets markedly greater use, it can undergo stress and struggle to maintain a semantic stability. Even within the area of language and mathematics education there are issues. Nowhere can this be seen more clearly than with the word "language" itself and even more with "discourse" and its cognate terms. There is an interesting phenomenon that can be seen at work with the words "geometry" and "algebra." In the nineteenth century, though at different times and places, these words became pluralized and moved to being count nouns: from "geometry" to "geometries" and so "a geometry," from "algebra" to "algebras" and so "an algebra." This linguistic shift marked a significant mathematical one. I want to point to "discourse" moving to "discourses" and "a discourse" as a shift of comparable form and, though likely not of comparable import to mathematics education, it is nonetheless worthy of attention. "Discourse" has quite a history: Shakespeare wrote of a character's discourse, and Descartes' appendix is *Discourse on Method*. My impression is that within mathematics education, "discourse" was at first simply a variant for "language," reflecting the challenge in finding a simple English word that evenhandedly includes both speech (or "talk") and writing (or "text"), though the adjective "discursive" at times has had more significance (e.g., "discursive practices" in Walkerdine, 1988). Now 'discourse' has taken on a life of its own, flowering widely and wildly in its usage.

A recent university library search (for a course I was teaching on Discourse Analysis in Education) produced fewer than ten book titles with the words "discourse analysis" in them pre-1975 and over a thousand by 2009. As mentioned by more than one author here, there is confusion, disagreement, and contention around the term, which does not make for clarity or great progress. Cue the need for Augean-stable activity. Part of the upcoming project for this area of research will be in stabilizing the sense and usage of this term.

IN CONCLUSION

[Art is the] making well of whatever needs making.
—Coomaraswamy, 1956, p. 91

In closing this opening rush of comment, I wish to point to the possibility that Truth in mathematics education not only needs making but well-making. And in the spirit of the call for greater interdisciplinarity contained in this book, I think the arts are an underutilized resource in this regard. There is great knowledge in practitioners of the arts about the fabrication of Truth (and I use the word "fabrication" advisedly—I want my Truth artfully well-made).

I have offered these glimpses as a few current instances I am aware of where classroom language explorations in mathematics classrooms are producing interesting phenomena that can throw some light on the more regular situation, one which is often harder to notice, and to highlight the dual track of working in this area, professional and academic, with its many crossovers, intersections, chicanes, and occasional collisions.

REFERENCES

Barton, B., Fairhall, U., & Trinick, T. (1998). *Tikanga reo tātai*: Issues in the development of a Māori mathematics register. *For the Learning of Mathematics, 18*(1), 2–9.

Barwell, R. (2009). *Multilingualism in mathematics classrooms: Global perspectives*. Bristol, UK: Multilingual Matters Press.

Coomaraswamy, A. (1956). *Christian and Oriental philosophy of art*. New York: Dover. (Originally published in 1943 with the title: *Why exhibit works of art?*)

Crespo, S. (2006). Examining prospective teachers' learning of three mathematics teaching practices—posing, interpreting, and responding—during teacher preparation. National Science Foundation NSF CAREER Grant No. 0546164.

Crespo, S., Oslund, J., & Parks, A. (2007). Studying elementary pre-service teachers' learning of mathematics teaching: Preliminary insights. In T. Lamberg & L. Wiest (Eds.), *Proceedings of the 29th annual meeting of the North American chapter of the International Group for the Psychology of Mathematics Education* (pp. 975–982). Stateline (Lake Tahoe), NV: University of Nevada, Reno.

Jamieson, K. (1975). Antecedent genre as historical constraint. *Quarterly Journal of Speech, 61*(4), 406–415.

Jefferson, G. (2004). Glossary of transcript symbols with an introduction. In G. Lerner (Ed.), *Conversation analysis: Studies from the first generation* (pp. 13–23). Philadelphia: John Benjamins.

Lakatos, I. (1976). *Proofs and refutations*. Cambridge: Cambridge University Press.

Love, E., & Pimm, D. (1996). "This is so": A text on texts. In A. Bishop (Ed.), *International handbook of mathematics education* (pp. 371–409). Dordrecht, Holland: Kluwer Academic Publishers.

Pimm, D. (2009). Method, certainty and trust across disciplinary boundaries. *ZDM – The International Journal on Mathematics Education, 41*(1–2), 155–159.

Pimm, D., Beisiegel, M., & Meglis, I. (2008). Would the real Imre Lakatos please stand up. *Interchange, 39*(4), 469–481.

Sabena, C., Radford, L., & Bardini, C. (2005). Synchronizing gestures, words and actions in pattern generalizations. In H. Chick & J. Vincent (Eds.), *Proceedings of the 29th conference of the International Group for the Psychology of Mathematics Education* (vol. 4, pp. 129–136). Melbourne: University of Melbourne.

Walkerdine, V. (1988). *The mastery of reason.* London: Routledge.

Zazkis, R., Liljedahl, P., & Sinclair, N. (2009). Lesson plays: Planning teaching versus teaching planning. *For the Learning of Mathematics, 29*(1), 40–47.

PREFACE

Judit N. Moschkovich

The inspiration for this volume came initially from Lauren Young of the Spencer Foundation, who recognized the need for a greater focus on the relationship between language and mathematics in order to support the diversity of students in today's classrooms. Initially, Lauren contacted each of the authors of chapters in this volume and requested that we write a position paper focused, broadly speaking, on the topic of "language and mathematics." She asked us to write papers that would help ground, define, and grow a future research agenda on issues of language as they relate to mathematics learning and teaching. Lauren's conversations with the authors of chapters in this book stimulated our thinking on the subject and encouraged us to develop the work presented here. We thank her and the Spencer Foundation for providing the motivation to write these pieces and for the support during the initial stages of writing.

As part of a series on research in mathematics education, this volume is principally intended for an audience of researchers, and graduate students in particular. Our hope is that new researchers will be able to use this volume as a starting point and grounding for future work on this topic. This volume is an interdisciplinary collection. While all the authors are scholars in the field of education research, several bring expertise in fields other than mathematics education. The bringing together of expertise in multiple fields was a purposeful choice. The topic of language and mathematics education requires not only interdisciplinary expertise but also interdisci-

Language and Mathematics Education, pages xiii–xvi
Copyright © 2010 by Information Age Publishing
All rights of reproduction in any form reserved.

plinary approaches and multiple perspectives on theory, research design, methods, and ways to apply research in classrooms.

As a researcher in mathematics education, I bring the lenses of the learning sciences and the field of mathematics education to the editing of this volume. My own research focuses on mathematical thinking, learning, and discourse. In order to address language issues in learning mathematics, I read across several sets of research literature. In doing this inter- and cross-disciplinary work, I found that, while I remained grounded in my own field, mathematics education, I was also using perspectives from fields in which I had little formal training, such as bilingualism and second language acquisition.

The multiple perspectives of the scholars in this volume who are not in mathematics education facilitate the interdisciplinary work required to address the topic of language and mathematics education. The authors not only provide important insights into the research they review, but they also provide the reader ways to conceptually organize research studies and to frame the central concepts distilled from research that are informed by their particular expertise. The reviews in this volume are neither comprehensive nor exhaustive and should not be treated as such. Each chapter provides an overview of only selected research literature. As a volume, the four chapters provide a broad view of the work relevant to the study of language and the learning and teaching of mathematics.[1] Below I describe the perspectives each chapter brings to the collection.

In Chapter 1, "Language(s) and Learning Mathematics: Resources, Challenges, and Issues for Research," I describe the theoretical notions about language that different sets of literature provide and some of the methodological issues that I have encountered. This chapter provides an introduction by describing resources, challenges, and methodological issues to consider when designing research on language and mathematics learning. To develop the necessary grounding to conduct this research, I read across disciplines and fields. I found that I faced several challenges and recurring issues, but I also encountered many useful resources. In this first chapter I describe the theoretical notions about language that different sets of literature provide and some of the methodological issues I considered.

Chapter 2, "Developing a Mathematical Vision: Mathematics as a Discursive and Embodied Practice," presents the combined perspectives of Kris D. Gutiérrez, Tesha Sengupta-Irving, and Jack Dieckmann. Gutiérrez brings a cultural historical activity theoretical approach to learning and development, particularly around language, literacy, and culture. Sengupta-Irving's work focuses on the social organization of mathematics classrooms as it relates to student participation, learning, and achievement. Dieckmann's work examines the interrelationship of language and mathematical practices, especially in classrooms with English learners at the secondary level.

In this chapter the authors examine major lines of inquiry in mathematics education through the prism of cultural historical activity theory, focusing on the language and discursive practices in the teaching and learning of school mathematics. In their review, the authors make a distinction between the language *in* and *of* mathematics learning in classrooms; use the notion of *mathematical vision* for mathematics learning that is robust, multi-semiotic, and embodied; and review work on everyday and scientific discourse practices. They revisit the theoretical principles from the work of Vygotsky, Engeström, and Cole, among others, to show how scientific or school-based mathematical learning "grows down into" the everyday and thus argue their relation.

In Chapter 3, "Language in Mathematics Teaching and Learning: A Research Review," linguist Mary Schleppegrell brings the lens of an applied linguistics perspective. She focuses on illuminating the contributions of linguists and applied linguists so that their work is better understood in the mathematics education community. Schleppegrell identifies key themes and major ideas in discussions of mathematics and language, focusing on central challenges and opportunities for work that would advance the field of mathematics education through a better understanding of the relationship between language and mathematics as it presents itself in mathematics classrooms. This chapter contributes to formulating a research agenda for understanding the linguistic challenges of teaching mathematics and preparing teachers for addressing these challenges.

In Chapter 4, "Function and Form in Research on Language and Mathematics Education," Guillermo Solano-Flores, formally trained in psychometrics, offers the perspective of a researcher in assessment conducting research informed by the language sciences (including sociolinguistics, cognitive science, second language acquisition, translation, and cultural anthropology, among other fields). Solano-Flores proposes that four views of language—as a process, system, structure, and factor—shape how researchers of different orientations have investigated language and its relation to mathematics, in particular when looking at testing practices. He suggests that in order to attain more valid and fair mathematics testing practices for linguistically diverse populations, researchers need to combine multiple views of language and communicate more effectively across areas of specialty.

In the final chapter, "Recommendations for Research on Language and Mathematics Education," I summarize the recommendations for future research on language and mathematics education proposed in the four chapters. These recommendations draw on multiple theoretical and methodological approaches and raise productive questions for further inquiry regarding language and mathematics learning, teaching, and assessment. I summarize central recommendations for future research, provide examples of research questions proposed in the four chapters as productive for future studies to pursue, and consider implications for instructional and assessment practices for English learners in mathematics classrooms.

Lastly, a word about the foreword and afterword is needed. The foreword by David Pimm provides a historical overview and personal account of the topic of language and mathematics education. As the author of a seminal book on this topic, *Speaking Mathematically: Communication in Mathematics Classrooms,* Pimm is well situated to provide these accounts. My own scholarly work on this topic was greatly influenced not only by that book but also by the interaction, support, and scholarly community that David and those scholars he has mentored have provided over the years.

The afterword provides a view by a scholar who works in learning sciences, focusing not on mathematics education but science education. Beth Warren's contribution is significant in at least two ways. The work of Cheche Konnen in language minority science classrooms should inform work in language issues in mathematics education. Researchers in mathematics education need to look to science education for work that is related to language. I give this advice from personal experience. Much of my own learning about language and discourse came from my short but amazingly productive time working with the Cheche Konnen team. I am grateful for all that I learned from the scholars on that team: Beth Warren, Ann Rosebery, Cindy Ballenger, and Josianne Houdicourt-Barnes.

I dedicate this book to the memory of my mother, Rosa Scheinbaum de Moschkovich. She taught me to love my own mother tongue (Spanish) by supporting my reading fiction (and not just doing physics and math, as I might have done if left to my own devices). She also taught me to love her mother tongue (Portuguese) and her mother's mother tongue (Idisch) by sharing her favorite songs with me. I am grateful to her for enriching my love of mathematics and physics with a passion for language.

I would like to thank the authors in this volume for their commitment and patience, and the series editor for her support. Last, but never least, I would like to thank my partner for being supportive and celebrating the results of my writing all these years since my very first published article.

—**Judit Moschkovich**
Santa Cruz, California

NOTE

1   The reviews have one limitation. The volume is largely focused on research literature in English from the United States and the United Kingdom.

CHAPTER 1

LANGUAGE(S) AND LEARNING MATHEMATICS

Resources, Challenges, and Issues for Research

Judit N. Moschkovich

ABSTRACT

My goals in this chapter are to describe resources, challenges, and methodological issues to consider when designing research on language and mathematics learning. As a researcher in mathematics education, I bring the lenses
of the learning sciences and the field of mathematics education to this project. Because my own research focuses on mathematical thinking, learning,
and discourse, both in monolingual and bilingual settings, I have had to read
across several sets of research literature. In doing this inter- and cross-disciplinary work, I found that, while I remained grounded in my own field, I was
using perspectives from fields in which I had little formal training, such as
bilingualism and second language acquisition. My personal experiences of
learning another language as a young child, being an immigrant, and becoming bilingual as an adolescent served to spark my curiosity about bilingualism
and second language acquisition. My commitment to improving the educa-

Language and Mathematics Education, pages 1–28

tion of learners who are from non-dominant groups has provided my motivation and has sustained my dedication to tackling these issues. To develop the
necessary grounding to conduct this research, I read across disciplines and
fields. I found that I faced several challenges and recurring issues, but I also
encountered many useful resources. In this chapter I describe the theoretical
notions about language that different sets of literature provide and some of
the methodological issues that I have encountered. Although I use examples
from the research literature, my goal was not to provide an exhaustive review
but to ground the chapter in examples.

INTRODUCTION

Integrating language into research on mathematics learning is an important goal for both practical and theoretical reasons. This integration is crucial for improving mathematics learning and teaching for students who are
bilingual, multilingual, or learning English. This integration is also relevant
to theory: Research with learners who use more than one language can
make language more visible than it might seem in monolingual situations,
providing a window on the role of language in learning mathematics. In
this chapter I describe resources, challenges, and methodological issues
to consider when designing research on the topic of language and mathematics learning. The first section focuses on the theoretical notions about
language that are provided by different sets of research literature and that
researchers should keep in mind if they address language. The second section describes several challenges to designing studies that are theoretically
framed and methodologically sound. The third section focuses on the theory and methods that are relevant for data collection and interpretation.

Addressing the relationship between language and mathematics learning
presents several challenges. The most significant challenge is that research
examining language and mathematics learning must be grounded not only
in current theoretical perspectives of mathematics cognition and learning,
but also in current views of language, classroom discourse, bilingualism,
and second language acquisition. Becoming familiar with two or more sets
of research literature can be a daunting task. The first section of this essay
describes resources available from several different fields to design research
on the topic of language and learning mathematics.

A second challenge in this endeavor arises because we all regularly participate in using language and, thus, we have developed intuitions about
language based on our personal experience. Our personal experiences
with language are couched in complex social, political, and historical contexts, and our intuitions may have developed into language attitudes. Our
intuitions about language may at times be in direct contradiction with
empirical research on how people acquire language or use two languages.

To address these contradictions, in the second section of the essay I describe common pitfalls to avoid when considering language in mathematics learning.

In the last section of the essay I describe important methodological issues to consider when designing research that addresses language(s) and mathematics learning, in particular when working with bilingual populations. This section focuses on data collection, transcription, and translation.

The greatest challenge is perhaps the term *language*. Many commentaries on the role of academic language in teaching practice reduce the meaning of the term language to single words and the proper use of grammar (for an example, see Cavanagh, 2005). In contrast, work on the language of specific disciplines provides a more complex view of mathematical language (e.g., Pimm, 1987) as not only specialized vocabulary (new words and new meanings for familiar words) but also as extended discourse that includes syntax and organization (Crowhurst, 1994), the mathematics register (Halliday, 1978), and Discourse practices (Moschkovich, 2007b). Theoretical positions in the research literature in mathematics education range from asserting that mathematics is a universal language, to claiming that mathematics is a language, to describing how mathematical language is a problem. Rather than joining in these arguments to consider whether mathematics is a language or reducing language to single words, I use a sociolinguistic framework to frame this essay. From this theoretical perspective, language is a socio-cultural-historical activity, not a thing that can either be mathematical or not, universal or not. I use the phrase "the language of mathematics" not to mean a list of vocabulary words or grammar rules but the communicative competence necessary and sufficient for competent participation in mathematical Discourse practices. I sometimes use the term "language(s)" to remind us that there is no pure unadulterated language and that all language is hybrid.

RESOURCES FOR DESIGNING RESEARCH ON LANGUAGE AND MATHEMATICS LEARNING

There are several sets of literature that can provide important theoretical notions when designing research studies on language and learning mathematics. Scholarly literature from several different fields, such as psycholinguistics studies on language switching during arithmetic calculations, sociolinguistics studies on code switching, research on classroom discourse in monolingual mathematics and science classrooms, and international research in multilingual classrooms, can contribute to designing further research.[1]

By necessity, researchers in mathematics education who address issues of language have used work from fields outside of mathematics education to inform research on the relationship between language and mathematics cognition and learning. Researchers in mathematics education have used work from other disciplines to examine the role of language in mathematics learning in bilingual classrooms in the U.S. (for example Khisty, 1995, 2001; Khisty & Chval, 2002; Moschkovich, 1999, 2002, 2007a) and in multilingual classrooms in South Africa (for example Adler, 1998, 2001; Setati, 1998; Setati & Adler, 2001).[2] Work outside of mathematics education has contributed theoretical frameworks for studying discourse in general, methodologies (e.g., Gee, 1996), concepts such as registers (Halliday, 1978) and Discourses (Gee, 1996), and empirical work on classroom discourse (e.g., Cazden, 1988; Mehan, 1979). Work in second language acquisition, bilingualism, and bi-literacy (e.g., Bialystok, 2001; Hakuta & Cancino, 1977; Valdés-Fallis, 1978, 1979; Zentella, 1997) has also provided theoretical frameworks, concepts, and empirical results that are relevant to this research endeavor. This work has provided crucial concepts necessary for studying the role of language in mathematics learning: for example, definitions of code switching (Auer, 1984; Gumperz, 1982; Zentella, 1981), distinctions among different types of code switching (Sanchez, 1994; Torres, 1997; Zentella, 1981), and the concept of hybridity (Gutierrez, Baquedano-Lopez, & Alvarez, 2001).

Borrowing concepts from other disciplines is a productive approach for integrating language into the study of mathematics learning. For example, the distinction between "national" languages—such as Spanish, English, or Haitian Creole—and "social" languages—such as mathematical or academic discourses—is useful in clarifying what we mean when we use the term "language." While concepts and theories from other disciplines provide essential resources, borrowing concepts also presents challenges. There is danger in borrowing concepts and leaving behind the intellectual tradition that gives a concept meaning. Those of us whose expertise lies in mathematics education must remember that notions such as language, culture, and bilingualism are as complex as any theoretical construct in our own field: These terms have contested meanings, long histories, and are the topics of heated debates in the fields of anthropology and linguistics.

In considering what work might be relevant to research on language and mathematics learning, it is important to distinguish between psycholinguistics and sociolinguistics because these two perspectives differ in how they conceptualize language. While sociolinguistics stresses the social nature of language and its use in varying contexts, psycholinguistic studies have been limited to an individual view of performance in experimental settings. From a sociolinguistic perspective, psycholinguistic experiments provide only limited knowledge about how people use language, because, as Hakuta and McLaughlin (1996) explain, "The speaker's competence is multifaceted: How a person uses language will depend on what is understood to be appropriate in a given social setting, and as such, linguistic knowledge is situated not in the individual psyche but in a group's collective linguistic norms" (p. 608).

"Bilingualism" is one example of a concept that has different meanings depending on the theoretical perspective one uses to frame it. The two perspectives described above see bilingualism differently. A researcher working from a psycholinguistic perspective might define a bilingual person as any individual who is in some way proficient in more than one language. This definition might include a native English speaker who has learned a second language in school with some level of proficiency but does not participate in a bilingual community. In contrast, a researcher working from a sociolinguistic perspective might define a bilingual person as someone who participates in multiple language communities and is "the product of a specific linguistic community that uses one of its languages for certain functions and the other for other functions or situations" (Valdés-Fallis, 1978, p. 4). This definition defines bilingualism not only as an individual but also as a social and cultural phenomenon that involves participation in language practices and communities.

An important resource for research addressing language and learning mathematics is research carried out in geographic settings with student populations other than the target population for a particular study. For example, researchers have studied language, bilingualism, and mathematics learning in Australia (e.g., Clarkson, 1991; Ellerton & Clements, 1991), Papua New Guinea (e.g., Clarkson, 1991; Clarkson & Galbraith, 1992; Dawe, 1983; Jones, 1982; Souviney, 1983), and in South African multilingual classrooms (e.g., Adler, 1998, 2001; Setati, 1998). This work can be an important resource for research with other student populations, as long as researchers note differences among settings that might be relevant to issues of language and learning mathematics for the student population for a particular research study.

What might be the relevance of work from Australia, Papua New Guinea, New Zealand, or the UK for Latino mathematics learners in the U.S.? What are the historical, political, and linguistic differences between the U.S. and South Africa that one should consider when using research from these two settings? Before applying research from Australia, New Zealand, PNG, or the UK to U.S. settings and student populations, researchers should carefully consider relevant differences among settings, students, languages, and communities. One difference is that the U.S. Latino population of school-age children can be largely described as bilingual in Spanish or as monolingual English speakers.[3] In contrast, the majority of students (as well as teachers) in South African classrooms speak multiple indigenous languages at home.

Another contrasting example is Pakistan, where the language of schooling is usually not spoken at home but reserved for activities related to school or government-related activities. Barwell (2003) provides some useful distinctions among different language settings, using the terms *monopolist, pluralist,* and *globalist.* In monopolist classrooms, all teaching and learning takes place in one dominant language; in pluralist classrooms, several languages used in the local community are also used for teaching and learning; in globalist classrooms, teaching and learning are conducted in an internationally used language that is not used in the surrounding community.

Another difference to consider across settings and national languages is the nature of the mathematics register in students' first language. For example, individual mathematical terms exist in Spanish. Since university-level courses and texts have existed in Spanish for centuries, the mathematics register in Spanish can be used to express many types of mathematical ideas from everyday to advanced academic mathematics. This may not be the case for the home languages of students in other settings such as South Africa (Setati & Adler, 2001) or in the case of Australasian Aboriginal languages and Maori (Barton, Fairhall, & Trinick, 1998; Roberts, 1998). These differences, however, should not be construed as a reflection of differences in learner's abilities to reason mathematically or to express mathematical ideas. Nor should we assume that there is a hierarchical relationship among languages that have different ways available to express school mathematical ideas. Instead, an ethno-mathematical perspective expands the kinds of activities considered mathematical beyond the mathematics found in textbooks or learned in schools (Bishop, 1986; D'Ambrosio, 1991; Nunes, Schliemann, & Carraher, 1993). This perspective emphasizes that mathematical activity is not a unitary category but is manifested in different ways in different settings, that all cultural groups generate mathematical ideas, and that Western mathematics is only one type of mathematical activity among many (Bishop, 1986). Taking an ethno-mathematical stance means that no mathematical activity is seen as a deviant, immature, or novice version of Western academic mathematical practices. Instead, mathematical activity is assumed to be situated as humans use social, cognitive, linguistic resources and cultural tools to make sense of problems.

In addition to ethno-mathematics, other research in mathematics education provides useful theoretical frameworks for integrating language into research on mathematics learning. As an example, two publications by Brenner provide theoretical distinctions that are relevant to both research and practice. Brenner (1994) provides useful distinctions among different kinds of communication in mathematics classrooms and describes three components of a "Communication Framework for Mathematics":

Communication About Mathematics entails the need for individuals to describe problem solving processes and their own thoughts about these processes.... Communication In Mathematics means using the language and symbols of mathematical conventions... Communication with mathematics refers to the uses of mathematics which empower students by enabling them to deal with meaningful problems. (p. 241)

The framework described by Brenner in another publication (1998) is useful for considering the relevance of research studies, organizing literature searches, and synthesizing across studies. Brenner provides a three-dimensional model for examining the cultural relevance of instruction and curriculum. This framework can be used to guide the design of research (as well as curriculum and instruction) by focusing research questions on aspects of classrooms that are relevant to three aspects of culturally relevant instruction: *cultural content, cognitive resources,* and *social organization.* As Brenner describes it, examining materials and instructional techniques for their *cultural content* can reveal the extent to which mathematical activities used in instruction relate to mathematical activities in local communities. Similarly, classrooms that make use of the *cognitive resources* students bring from previous instruction and from home—a variety of ways of thinking used to solve problems—make the most of students' existing knowledge. And lastly, ensuring that classroom *social organization* facilitates comfortable and productive participation for all students and takes into account a variety of possible roles, responsibilities, and communication styles will more likely support comfortable and productive student participation.

The first and second aspects of mathematics instruction, *cultural content* and *cognitive resources,* have been addressed by research that examines the mathematical activities in local communities (e.g., González, Andrade, Civil, & Moll, 2001; Moll, Amanti, Neff, & González, 1992) and the cognitive resources students bring to the classroom for learning mathematics (for example, algorithms across cultures; see Orey, 2003; Secada, 1983). The third aspect, *social organization,* seems most relevant to issues of language. What constitutes comfortable and productive participation for different populations of mathematics learners who are bilingual, multilingual, and/or learning English? Sociolinguistic studies that examine how young bilingual learners use two languages (e.g., Zentella, 1997) provide a relevant knowledge base to address this question. What cultural models for communication do students bring to the mathematics classroom? This question might be approached using research studies that have examined models for communication among particular student populations—for example, native Hawaiian children (Au, 1980; Au & Jordan, 1981), Navaho students (Vogt, Jordan, & Tharp, 1987), and African-American children (Heath, 1983; Lee, 1993). Empirical research on communication models for other student populations in the U.S. should also provide a relevant knowledge base for further research. However, research on communication models should be used with caution, as examples of how communication practices can vary, but not to make broad generalizations about the communication styles for a particular group of learners or an individual.

Many authors have warned researchers repeatedly about the danger in assuming that communication styles or home cultural practices are homogeneous in any community, dominant or non-dominant. For example, Gutierrez, Baquedano-Lopez, and Alvarez (2001) describe language practices as "hybrid" and based on more than one language, dialect, or practice. Gutiérrez and Rogoff (2003) also caution us against ascribing cultural practices to individuals and instead propose we consider the repertoires of practices that any one individual has had access to. We cannot assume that *any* cultural group has "cultural uniformity or a set of harmonious and homogeneous shared practices" (González, 1995, p. 237). González (1995) decries perspectives that "have relegated notions of culture to observable surface markers of folklore, assuming that all members of a particular group share a normative, bounded, and integrated view of their culture" and suggests that "approaches to culture that take into account multiple perspectives can reorient educators to consider the everyday experiences of their student" (p. 237). Still, Brenner's three-part framework can be used as a guide to consider more broadly the complexity in what might constitute comfortable and productive participation for learners as well as the varied practices that students have experienced in multiple settings. Researchers should keep in mind that learners from any community can and do participate productively in a variety of roles, responsibilities, communication styles, and mathematical activity that include hybrid practices.

CHALLENGES IN DESIGNING RESEARCH ON LANGUAGE AND MATHEMATICS LEARNING

Examining language and learning mathematics presents many interesting challenges. Barwell (2003) has described what he calls "linguistic discrimination," giving examples of how research studies sometimes assume homogeneity, position English as the norm and default language, or use written tests as the sole instruments to describe linguistic competence. In the next section I describe several other pervasive and enduring challenges: defining bilingualism, building on previous work, avoiding deficit models of bilingual learners, and avoiding superficial conclusions about language and cognition.

Defining Bilingualism

The first challenge researchers face when designing research with bilingual/multilingual populations is how these labels are used in ambiguous ways and with multiple meanings. Research studies need to be clear in specifying how the labels *bilingual* or *multilingual* are used, for both classrooms and learners. There are many different labels for different types of bilingual classrooms, and these labels do not always tell us what, exactly, happens in the classroom in terms of how teachers and students use each of the languages they may also use outside the classroom.

Studies should document and report not only students' proficiency in each language but also their histories, practices, and experiences with each language across a range of settings and tasks. Studies should describe proficiencies in each language wherever possible in both oral and written modes. There are serious challenges that such research will need to address, such as the complexity of defining language "proficiency," the lack of instruments that are sensitive to oral and written modes, and the scarcity of instruments that address features of the mathematics register for specific mathematical topics.

Studies should not assess English or Spanish proficiency *in general* but rather specifically for communicating in writing and orally about a particular mathematical topic. Students have different opportunities to talk and write about different mathematical topics in each language, in informal and instructional settings. Assessments, then, should consider not only proficiency in each language but also proficiency for using each language to talk or write about a particular mathematical topic. One example of how researchers might approach this challenge is Secada's (1991) study that used a complex view of language proficiency and examined the semantic structure of arithmetic word problems. Language proficiency was assessed using the Language Assessment Scales (De Avila & Duncan, 1981b; Duncan & De Avila, 1986, 1987), oral story telling, and verbal counting up and down. The instruments assessed syntax, phonetics, lexicon, and pragmatics and included language tasks that were closely related to the specific mathematical thinking and topic examined in the study.

Studies should also document and report students' mathematical histories, practices, and experiences across a range of settings and tasks involving mathematics. It is important to examine mathematics learning in the context of its development while considering the tools and practices employed for mathematical reasoning. It is crucial that in documenting mathematical practices, researchers consider the spectrum of mathematical activity as a continuum rather than reifying the separation between practices in and out of school. Lastly, research should consider mathematical activity from an emic perspective, as it is experienced by the participants rather than imposing the view of a mathematical expert on the analysis of these practices. Neglecting these three aspects of mathematical practices typically leads to a deficit view of everyday mathematical practices.

Definitions of bilingualism range from native-like fluency in two languages to alternating use of two languages (De Avila & Duncan, 1981a) to belonging to a bilingual community (Valdés-Fallis, 1978). In my own work, I use the definition of bilinguals provided by Valdés-Fallis as "the product of a specific linguistic community that uses one of its languages for certain functions and the other for other functions or situations" (p. 4). This definition characterizes bilingualism as not only an individual but also a social and cultural phenomenon involving participation in language practices and communities.

A common misunderstanding of bilingualism is the assumption that bilinguals are equally fluent in their two languages. If they are not, then they have been described as not truly bilingual and sometimes labeled as "semilingual" or "limited bilingual." In contrast, current scholars studying bilingualism see "native-like control of two or more languages" as an unrealistic definition that does not reflect evidence that the majority of bilinguals are rarely equally fluent in both languages (Grosjean, 1999). For example, Grosjean proposes we shift from using the terms "monolingual" and "bilingual" as labels for *individuals* to using these as labels for the endpoints on a continuum of *modes*. Researchers have also strongly criticized the concept of semilingualism and proposed that we leave that notion behind (for a review of criticisms, see Baetens Beardsmore, 1986; MacSwan, 2000).

One solution to the use of vague and confusing labels is for researchers to know the students and the community. In this way, they can make careful and informed decisions regarding what questions to ask in order to be able to describe the cultural and linguistic context clearly and in detail. There are many questions to ask about mathematics learners, and these questions have complex answers: Do students themselves identify as bilingual or monolingual? Does the school identify students as monolingual or bilingual? If so, on what basis are students labeled bilingual or monolingual? Is a student a recent immigrant, first generation, second generation, or part of borderland communities? Do students come from rural, urban, migrant, or farming settings? What are students' previous schooling experiences? How many years have they been in school? How much mathematics instruction have students experienced in each language? What are students' informal mathematical experiences with activities such as selling and buying, games, or work-related mathematics?

Avoiding Deficit Models of Learners and Their Communities

A crucial pitfall to avoid when examining language and mathematics learning for students who are bilingual, multilingual, or learning English is using deficit models of language minority learners and their communities. Many deficit models stem from assumptions about learners and their communities based on race, ethnicity, SES (socio-economic status), and other characteristics assumed to be related in simple, and typically negative, ways to cognition and learning in general. For example, U.S. policy and research, rather than seeing bilingual Latino learners as having additional language skills, have used deficit models to describe these students (Garcia & Gonzalez, 1995).

Deficit models are so pervasive and insidious that we can sometimes fail to recognize them. For example, any time we use monolingual learners (or classrooms) as the norm, we are imposing a deficit model on bilingual learners. Bilinguals learning mathematics need to be described and understood on their own terms and not only by comparison to monolinguals. Future research should move away from comparisons between monolingual and bilingual learners. Studies should focus less on comparisons to monolinguals and report not only differences between monolinguals and bilinguals but also similarities. Studies focused on the differences between bilingual and monolinguals may miss the similarities, for example, in how both types of students solve mathematics problems.

Early research focused on the disadvantages that English learners and bilinguals face, focusing on comparing response times between monolinguals and bilinguals (McLain & Huang, 1982; Marsh & Maki, 1976) or the obstacles the mathematics register in English presents for English learners (Spanos & Crandall, 1990; Spanos, Rhodes, Dale, & Crandall, 1988). Research has not yet seriously considered any possible advantages of bilingualism for mathematics learning. Future research should stop focusing on the disadvantages associated with learning English or being bilingual and explore any advantages that bilingualism might provide for learning mathematics. One example of an advantage for bilinguals is the reported role of attention in solving mathematical problems. After reviewing research on the cognitive consequences of bilingualism, Bialystok (2001) concluded that bilinguals develop an "enhanced ability to selectively attend to information and inhibit misleading cues" (p. 245).[4] This conclusion is based, in part, on the advantage reported in one study that included a proportional reasoning task (Bialystok & Majumder, 1998) and another using a sorting and classification task (Bialystok, 1999). Although these tasks seem closely related to mathematical problem solving, they have not been examined in detail in the context of bilinguals doing or learning mathematics.

Another common pitfall to avoid is blaming the parents, the community, or the culture for perceived deficits in the learners. This is a complex issue, best illustrated by an example. On the one hand, research has shown that parental education is related to student performance. For example, a study of 4th and 8th grade LEP (Limited English Proficiency) and non-LEP students in the U.S. found that students of parents with less than a high school education had lower average reading and math scores (Abedi, Leon, & Mirocha, 2003). Some LEPs with highly educated parents had higher scores than non-LEPs with parents with less than a high school education. However, it would be a simplistic application of a deficit model of the community to then conclude that the parents are the problem and that parents with less than a high school education perhaps do not care about education or do not help students with homework. In the case of Latino parents in the U.S., empirical research shows that this is simply not the case. In a national survey of Latinos on the topic of education, more Latino parents reported that they attended PTA meetings than non-Latino white parents and about the same number of Latino parents reported that they regularly helped their children with homework on a daily basis (Pew Hispanic Center, 2004). A similar percentage of Latino and Anglo students reported parents regularly reviewed their homework (National Center for Educational Statistics [NCES], 1995). Latino 8th graders were more likely than their Anglo counterparts to report that parents had limited their TV viewing and that parents had visited their classes (NCES, 1995). Lastly, but perhaps most importantly, ethnographic studies repeatedly show that immigrant and Latino parents do, in fact, value education.

Focusing on the mathematical activity is another important way to avoid using deficiency models of bilingual learners. If research examining bilingual learners does not focus on the mathematical activity, then it may seem that bilingual learners do not engage in mathematical activity, thus further contributing to seeing them as deficient. It is crucial to uncover the mathematics bilingual learners are doing or are capable of doing. If analyses in mathematics education do not focus on the mathematical activity, then we contribute to a view that these learners are not *really* doing mathematics. Mathematics education research should keep the focus on the mathematical ideas and bring out the mathematics that bilingual and multilingual learners are engaged in.

In order to focus on the mathematical meanings learners construct, rather than the mistakes they make, researchers will need frameworks for recognizing the mathematical knowledge, ideas, and learning that learners are constructing in, through, and with language. A functional theory of language such as functional systemic linguistics (e.g., O'Halloran, 1999; Schleppegrell, 2007), a communication framework for mathematics instruction (Brenner, 1994), or a sociocultural perspective can serve as frameworks for recognizing mathematical contributions by students.

A sociocultural perspective on bilingual mathematics learners (Moschkovich, 2002) shifts the focus from looking for deficits to identifying the mathematical discourse practices evident in student contributions (e.g., Moschkovich, 1999). The sociocultural perspective (Moschkovich, 2002, 2004, 2007b) also provides a theoretical framework for recognizing the mathematics in student contributions. This framework assumes that mathematical Discourse is complex, grounded in practices, and connected to mathematical concepts. I use the phrase "Discourse practices" to emphasize that Discourse are not individual, static, or refer only to language. Instead, I assume that meanings are multiple, situated, and connected to communities. Discourses involve more than language; they also involve other symbolic expressions, objects, and communities. Discourse practices involve not only language but also perspectives and conceptual knowledge. Words, utterances, and texts have different meanings, functions, and goals depending on the practices in which they are embedded.[5] Discourses occur in the context of practices and practices are tied to communities. Discourse practices are constituted by actions, meanings for utterances, focus of attention, and goals, and these actions, meanings, focus, and goals are embedded in practices.[6]

Avoiding Superficial Conclusions About Language and Mathematical Cognition

Another important pitfall to avoid is jumping to conclusions regarding language and cognition. Below I discuss two common conclusions about bilinguals and mathematical thinking made on the basis of observations of two common practices among bilingual mathematics learners, using two languages during mathematical conversations or when carrying out arithmetic computation.

One common practice among bilinguals is switching languages during a conversation, a phenomenon called code switching. It is crucial to avoid reaching superficial conclusions regarding code switching and cognition— for example, that bilinguals switch languages because they do not remember a word or do not know a concept. Regardless of what our personal experiences or folk explanations of code switching may tell us, empirical research in sociolinguistics has shown that code switching is a complex language practice and not evidence of deficiencies. In general, code switching is not primarily a reflection of language proficiency, discourse proficiency, or the ability to recall (Valdés-Fallis, 1978). Bilinguals use the two codes differently depending on the interlocutor, domain, topic, role, and function.

For example, young bilinguals (beyond age 5) "speak as they are spoken to" (Zentella, 1981). Last but not least, when using two codes, the choice of one rather than the other involves an act of identity. Therefore it is not warranted to make simple conclusions about someone's code switching and their proficiency in a national language or in mathematics.

Another common practice among adult bilinguals is carrying out arithmetic computation in their first language and then translating the answer. A small set of psycholinguistic studies that have explored this phenomenon are based on the hypothesis that since bilinguals tend to do mental arithmetic in their first language, they therefore take longer to compute. An example of this work comes from two studies conducted with adult US Spanish speakers (McLain & Huang, 1982; Marsh & Maki, 1976). Marsh and Maki (1976) found that adult bilinguals performed arithmetic operations more rapidly in their preferred language than in their non-preferred language. The subjects (20 adult immigrants in the U.S. who preferred either English or Spanish) reported that they performed arithmetic in their preferred language and then translated answers to the non-preferred language. The difference between performance in a preferred language and a non-preferred language was slight (200 milliseconds). Comparisons between monolinguals and bilinguals (who preferred English to Spanish) showed a slight but statistically significant difference of about .5 seconds for mean response time.

The evidence regarding bilinguals' speed during arithmetic computation is inconclusive and contradictory. All we can safely say at this time is that "retrieval times for arithmetic facts *may* be slower for bilinguals than monolinguals" (Bialystok, 2001, p. 203). This possible difference is incredibly small (on the order of .5 seconds) and has been documented clearly only with Spanish and Japanese speaking adults, not young children. It is not clear whether these reported small differences in response time between monolinguals and bilinguals exist for young learners.

Overall, there is strong evidence suggesting that bilingualism does not impact mathematical reasoning. In the words of one researcher:

> The most generous interpretation that is consistent with the data is that bilingualism has no effect on mathematical problem solving, providing that language proficiency is at least adequate for understanding the problem. Even solutions in the weaker language are unhampered under certain conditions. (Bialystok, 2001, p. 203)

Balancing Building on and Revising Previous Work

Another challenge we face when designing research on language and learning mathematics is finding ways to build on older work and thus avoid reinventing wheels, while challenging and revising older work in light of new theories and data. We need to revisit early work in light of current research and also challenge outdated approaches when this is necessary. For example, early work on language and learning mathematics focused on how the mathematics register presented an obstacle for English learners (e.g., Spanos & Crandall, 1990; Spanos et al., 1988) and did not examine any of the resources English learners bring. In contrast, more recent work considers not only the obstacles for learners of a second language, but also the resources learners bring to the task of learning mathematics in a second language (e.g., Moschkovich, 2000, 2002).

Cummins' threshold hypothesis (1979) and the concept of "semilingualism" are examples of early work that should be revisited. The notion of semilingualism arose, in part, from an obsession with monolingualism as a norm. This notion is controversial, and should be used with great caution (Baetens Beardsmore, 1986). Semilingualism was first introduced by Nils Erik Hansegard (a Swedish philologist) in 1962 (in the absence of any theory of language) to conjecture that a period of "double semilingualism" occurs when an individual abandons her native language altogether in favor of a second language (MacSwan, 2000). In the U.S., Cummins (1976) used the term in describing the *Threshold Hypothesis*, which theorized that the level of linguistic competence attained by a bilingual child in a first and second language may affect his or her cognitive growth in other domains. Cummins defined "semilingualism" as low level in both languages (Cummins, 1979) to describe students who do not develop "native-like competence in either of their two languages" (Cummins, 1976, p. 20). This definition involves the conjecture that some children have limited or nonnative ability in the language or languages they speak (MacSwan, 2000).

Currently, most scholars have discarded the concept, even Skutnabb-Kangas and Cummins, two of the early proponents of this notion: "I do not consider semilingualism to be a linguistic or scientific concept at all. In my view, it is a political concept" (Skutnabb-Kangas, 1984, p. 248); "There appears to be little justification for continued use of the term semilingualism in that it has no theoretical value and confuses rather than clarifies the issue" (Cummins, 1994, p. 3813). The grounds for discarding the concept range from its nebulous nature to a lack of empirical support and theoretical foundation for the notion. "Semilingualism does not exist, or put in a way which is non-refutable, has never been empirically demonstrated" (Paulston, 1982, p. 54).

Critics of the concept of "semilingualism" argue that it confounds language proficiency (or linguistic competency) and use of academic register, formal schooling, SES, or "language loss" (the shift in choice of language occurring across generations). The concept also confuses degrees of ability, levels of linguistic competence, or levels of language development with differences in experience with language varieties (dialects, registers, and discourses) or with school literacy (reading, writing, and other aspects of language use valued in school).

Perhaps the strongest argument against semilingualism is the empirical evidence that it is not possible to have limited or nonnative ability in the language of one's home community. Linguists agree that "all normal children acquire the language of their speech community with some minor but ordinary degree of variation" and that "a native language is acquired[7] effortlessly and without instruction by all normal children" (MacSwan, 2000, p. 25).[8]

METHODOLOGICAL ISSUES IN DESIGNING RESEARCH ON LANGUAGE AND LEARNING MATHEMATICS

While the term "methodology" is sometimes used to refer only to "methods," I use the term "methodology" to refer to theory and methods together because theory and methods are mutually constructive (Moschkovich & Brenner, 2000). I assume that methodology includes the underlying theoretical assumptions about cognition and learning: what cognition and learning are; when and where cognition and learning occur; and how to document, describe, and explain these phenomena. All the issues raised in this section pertain not only to the methods one uses to examine language and mathematics learning but, more fundamentally, to the theoretical way that one frames and conceives of both mathematical activity and language.

There are methodological issues that are specific to designing and conducting research on language and learning mathematics. Most importantly, research needs to address and focus on the needs of specific student populations rather than using categories such as English learners, bilinguals, or Latinos as if these groups were homogeneous. "English learner" is a particularly vague label that does not capture the linguistic complexity of student experiences when learning mathematics. This category can include multiple student populations and collects learners into one category even though they have different experiences and needs—for example, Latinos who are monolingual, Latinos who are bilingual, Asian students, immigrants, young children, adolescents, and so on.

Data Collection

When designing research on language and learning mathematics, it is important to consider what data to collect, which tools to use and how to assess language proficiency. In general, the design of data collection should consider and build on the instruments used in previous research literature that is relevant, such as assessments of language proficiency in a particular topic in Spanish (for example, whole number operations in Secada, 1991) or assessments of reading proficiency in English that use traditional word problems (e.g., Clarkson & Galbraith, 1992). The analysis of classroom activity should be couched in and framed by many types of data: the teacher's goals, textbook use, district policies, preceding lessons, information about the students, etc. Contextualizing data is important, especially for cross-cultural work (Erickson, 1986). "Similar behaviors may have different meanings and comparisons can be problematic" (Ulewicz & Beatty, 2001, p. 13).

Research studies need to provide detailed ethnographic details about students' language backgrounds and experiences with mathematics in and out of school. Important information to collect when working with bilingual populations includes whether students are immigrants, first generation, or second generation. Information about previous schooling experiences is also important, such as whether students have participated in mathematics classes in their first language or not. Information on previous mathematical experiences outside of school, such as selling, buying, games, and so on, is also relevant. Research should document and report not only information about students' proficiency in each language but also their experiences with each language and with mathematics at home and at school. Assessing language proficiency is complex and involves oral, written, listening, and academic registers in content areas. Studies should distinguish between proficiencies in oral and written modes and, wherever possible, describe proficiencies in each language in more than one mode (oral and written).

It seems especially important to consider *how* we might use video data to examine language, mathematical activity, and mathematics learning with bilingual student populations, particularly for evaluative analyses of student activity. It is a common experience when analyzing video data to focus on what a student is doing wrong rather than on what a student is doing well. Because video slows action down, participants on videotape may seem both less and more competent than in real time. On one hand, as we watch video we have more time to notice how participants misspeak or make mistakes than we would have if we were observing in real time, thus making them appear *less* competent. This tendency to focus on the negative when looking at videotapes is a special danger when we analyze bilingual, multilingual, or English learners' mathematical activity. When looking at video data of bilingual or multilingual learners, it is especially important not to equate a participant's linguistic competence with competence in mathematical reasoning. On the other hand, as we watch video we also have more time to notice and really think about what participants said and did, potentially making them look *more* competent than in real time. Thus, video data also opens up the possibility to document student competence in mathematical reasoning.

Transcription and Translation

There are several challenges involved in working with more than one national language. Two important issues for research that integrates issues of language and mathematics learning are transcriptions and translations. Transcription and transcript quality are theory laden (Ochs, 1979; Poland 2002). Researchers make many decisions about transcripts that are based on their theoretical framework and on the particular research questions for a study. For example, decisions regarding what to include in transcripts and which transcript conventions to use are informed by theory. Whether a transcript will include gestures, emotions, inscriptions, body posture, and description of the scene (Hall, 2000; McDermott, Gospodinoff, & Aron, 1978; Poland, 2002) will depend on whether these aspects of activity are relevant or not to the particular research questions. Similarly, selecting transcript conventions and deciding whether overlapping utterances, intonation, and pauses are included or not in a transcript depends on whether these aspects are relevant to the research questions and analysis that the transcript and video will be used for, and whether and how aspects of activity are relevant (or not) to the research questions depends on the theoretical framework.

Two aspects of bilingual conversations and features of talk that may be relevant to documenting mathematics learning are intonation and the use of gestures. Perceiving a student as uncertain or hesitant because of intonation patterns may have an impact on how researchers (and teachers) perceive student contributions in mathematics classrooms. For example, intonation patterns vary across languages and among dialects. "Perhaps the most prominent feature distinguishing Chicano English from other varieties of American English is its use of certain intonation patterns. These intonation patterns often strike other English speakers as uncertain or hesitant" (Finegan & Besnier, 1989, p. 407).

Bilingual students' use of gestures to convey mathematical meaning has been documented (e.g., Moschkovich 1999, 2002). Further exploration of the use of gestures during mathematical discussions would provide more detailed descriptions of the role of gestures and intonation in how bilingual/multilingual mathematics learners communicate.

Translation presents a challenge all its own. Translation is not simply a copy of the original utterance since translating involves interpretation. It is impossible to translate without putting some piece of ourselves in the new utterance—translators are not simply empty vessels. When participants use two languages, it is important for researchers to choose clear ways to display transcripts and decide how the transcript will show translations. It is imperative that both the actual utterances of the participants in one (or more) language(s) as well as the translations be included in presentations and publications reporting on the research. Subtitles can be a useful way to display both an utterance and its translation, allowing other researchers to inspect both simultaneously.

Needless to say, translations should be done as carefully as transcriptions. Transcripts in another language should be as accurate as English transcripts, using not only the skills of native speakers but also professional translators. Knowing the particular community of students and checking translations with native speakers of the particular regional dialect are crucial aspects of translation work. Translators need to pay attention to the way that language is used in each particular community. Ideally, translators need to have knowledge of regional and national variation and know the community and the students. For example, transcribers and translators need to be knowledgeable about the particular variety of Spanish used (for example, Puerto Rican or Mexican-American) so that words, phrases, and expressions particular to each community are accurately translated. Researchers who do not speak the second language fluently themselves will need to be aware of whether a transcriber or translator is fluent in particular dialects of that language and has experience using the mathematics register. An important question to consider for translations of mathematical activity in more than one language is how domain knowledge impacts translation. Ideally, translators need to have mathematical knowledge and know mathematical terms and expressions in both languages. Involving a professional translator and using a second translator to check are useful strategies for managing the many challenges of translating children's mathematical discussions in two languages.

CLOSING

In closing, I describe a few general categories of research questions that seem fruitful for future research to consider. In order to be able to design instruction that builds on student resources, research needs to examine in more detail the resources that bilingual or English learning students use for mathematical reasoning. Many more studies are needed that describe how students who speak more than one language use multiple resources such as two languages, gestures, objects, and mathematical representations or inscriptions to communicate mathematically. Studies will need to distinguish among multiple modalities (written and oral) as well as between receptive and productive skills. Other important distinctions are between listening and oral comprehension, comprehending and producing oral contributions, and comprehending and producing written text.

It is also important for research to move away from construing everyday and school mathematical practices as a dichotomous distinction. During mathematical discussions students use multiple resources from their experiences across multiple settings, both in and out of school. Everyday practices should not be seen only as obstacles to participation in academic mathematical Discourse. The origin of some mathematical Discourse practices may be everyday practices, and some aspects of everyday experiences can provide resources in the mathematics classroom. Everyday experiences with natural phenomena can be resources for communicating mathematically. For example, climbing hills is an experience that can be a resource for describing the steepness of lines (Moschkovich, 1996). Other everyday experiences with natural phenomena may provide resources for communicating mathematically.

In addition to experiences with natural phenomena, O'Connor (1999) proposes that students' mathematical arguments can be at least partly based on what she calls argument protoforms: "Experiential precursors (arguments outside of school, the provision of justification to parents and siblings, the struggle to name roles or objects in play) may provide the discourse 'protoforms' that students could potentially build upon in the mathematical domain" (p. 27). These precursors are related to academic mathematical Discourse practices such as arguing, making and evaluating a claim, providing justification, or co-constructing a definition. Research should consider what aspects of everyday discourse could serve as resources for mathematical arguments.

Academic English (Scarcella, 2003) for mathematical communication is another important topic for research studies to address. Studies need to examine in more detail what exactly constitutes academic English competency for mathematics in both written and oral modes. If it is the case that academic English is different for different mathematical domains or genres of mathematical texts, then these differences need to be examined. Studies also need to explore how immigrant students transition from learning mathematics in their native language to learning mathematics in English: What are students' experiences learning mathematics in their first, second, and both languages? How do different proficiencies in a first language (oral, reading, written, academic English) and previous mathematics instruction in a first language impact students' learning mathematics in English?

Learning to read and use vocabulary in mathematics are two topics that also need attention from researchers. Studies are needed that examine how English learners learn to read different mathematical texts (textbooks, word problems, etc.). In designing this research it will be important to differentiate between reading textbooks and reading word problems, two different genres in mathematical written discourse. When working with children who are learning to read in English, it will also be important to distinguish between children who are competent readers in a first language and those children who are not. Research also needs to consider what are successful ways for English learners to learn vocabulary in mathematics. This work will need to start by establishing what vocabulary assessment instruments are relevant to English learners or bilingual students learning mathematics.

The question is not whether students should learn vocabulary but rather how instruction can best support students learning both vocabulary and mathematics. Vocabulary drill and practice is not the most effective instructional practice for learning either vocabulary or mathematics. Instead, vocabulary and second language acquisition experts describe vocabulary acquisition in a first or second language as occurring most successfully in instructional contexts that are language rich, actively involve students in using language, require both receptive and expressive understanding, and require students to use words in multiple ways over extended periods of time (Blachowicz & Fisher, 2000; Pressley, 2000). We already know that to develop written and oral communication skills, students need to participate in negotiating meaning (Savignon, 1991) and in tasks that require output from students (Swain, 2001). Researchers in vocabulary acquisition agree that the best way for students to develop mathematical vocabulary is have opportunities provided for them to actively use mathematical language to communicate about and negotiate meaning for mathematical situations. It is especially important that instruction for this population not emphasize low-level language skills over opportunities to actively and repeatedly communicate about mathematical ideas. One of the goals of mathematics instruction for bilingual students should be to support all students, regardless of their proficiency in English, in participating in discussions that focus on important mathematical ideas, rather than on pronunciation, vocabulary, or low-level linguistic skills.

NOTES

1. My intention here is not to provide an exhaustive literature review, but rather to mention a few selected and relevant studies. More extensive, exhaustive, and detailed reviews of the literature can be found in Adler (2001), Clarkson (1991), Ellerton & Clements (1991), and Moschkovich (2002, 2007c).
2. Because this chapter focuses on children learning mathematics, I do not include here research documenting teachers who are code switching in classrooms. For examples, see Khisty (1995), Setati (1998), Setati & Adler (2001), Valdés-Fallis (1978), and Zentella (1981).

3. There are Latino children and adults in the U.S. who also speak an indigenous language as their first language, Spanish as a second language, and English as a third language.
4. Bialystok (2001) mentions that the cognitive advantages of bilingualism seem to depend on some level of proficiency in both languages.
5. I use the terms *practice* and *practices* in the sense used by Scribner (1984) for a practice account of literacy to "highlight the culturally organized nature of significant literacy activities and their conceptual kinship to other culturally organized activities involving different technologies and symbol systems" (p. 13).
6. For a description of how discourse practices involve actions and goals and an analysis of the role of goals in the appropriation of mathematical practices, see Moschkovich (2004). For an analysis of how meanings for utterances reflect particular ways to focus attention, see Moschkovich (2008).
7. The notion of acquiring a language does not imply that once one acquires the language of one's community, this is a static end in language learning. The discussion throughout this essay should make it clear that language acquisition and proficiency are nuanced, situated, and dynamic processes.
8. While setting aside the notion of semilingualism, some researchers agree that variation exists in students' proficiency in educationally relevant aspects of language, in the formal language skills in one or more languages, and that bilingual learners need to develop what is currently called "academic English." The topic of academic English, however, is beyond the scope of this essay (for a discussion of academic English, see Cummins, 2000).

REFERENCES

Abedi, J., Leon, S., & Mirocha, J. (2003). *Impact of student language background on content-based performance: Analyses of extant data* (CSE Tech. Rep. No. 603). Los Angeles: University of California, National Center for Research on Evaluation, Standards, and Student Testing.

Adler, J. (1998). A language of teaching dilemmas: Unlocking the complex multilingual secondary mathematics classroom. *For the Learning of Mathematics, 18*(1), 24–33.

Adler, J. (2001). *Teaching mathematics in multilingual classrooms.* Dordrecht, The Netherlands: Kluwer Academic Press.

Au, K. (1980). Participation structures in a reading lesson with Hawaiian children: Analysis of a culturally appropriate instructional event. *Anthropology and Education Quarterly, 11,* 91–115.

Au, K., & Jordan, C. (1981). Teaching reading to Hawaiian children: Finding a culturally appropriate solution. In H. Trueba, G. Guthrie, & K. Au (Eds.), *Culture and the bilingual classrooms: Studies in classroom ethnography* (pp. 139–152). Rowley, MA: Newbury House.

Auer, P. (1984). *Bilingual conversation.* Amsterdam: John Benjamins.

Baetens Beardsmore, H. (1986). *Bilingualism: Basic principles.* Clevedon: Multilingual Matters.

Barton, B., Fairhall, U., & Trinick, T. (1998). Tikanga Reo Tātai: Issues in the development of a Māori mathematics register. *For the Learning of Mathematics, 18*(1), 3–9.

Barwell, R. (2003). Linguistic discrimination: An issue for research in mathematics education. *For the Learning of Mathematics, 23*(2), 37–43.

Bialystok, E. (1999). Cognitive complexity and attentional control in the bilingual mind. *Child Development, 70,* 636–644.

Bialystok, E. (2001). *Bilingualism in development: Language, literacy and cognition.* Cambridge: Cambridge University Press.

Bialystok, E., & Majumder, S. (1998). The relationship between bilingualism and the development of cognitive processes in problem solving. *Applied Psycholinguistics, 19,* 69–85.

Bishop, A. (1986). Mathematics education in its cultural context. *Educational Studies in Mathematics, 10*(2), 135–146.

Blachowicz, C., & Fisher, P. (2000). Vocabulary instruction. In M. Kamil, P. Mosenthal, P. D. Pearson, & R. Barr (Eds.), *Handbook of Reading Research* (vol. III, pp. 503–523). Mahwah, NJ: Lawrence Erlbaum Associates.

Brenner, M. (1994). A communication framework for mathematics: Exemplary instruction for culturally and linguistically diverse students. In B. McLeod (Ed.), *Language and learning: Educating linguistically diverse students* (pp. 233–268). Albany: SUNY Press.

Brenner, M. (1998). Adding cognition to the formula for culturally relevant instruction in mathematics. *Anthropology & Education Quarterly, 29*(2), 214–244.

Cavanagh, S. (2005). Math: The not-so-universal language. *Education Week,* July. http://www.barrow.k12.ga.us/esol/Math_The_Not_So_Universal_Language.pdf.

Cazden C. (1988). *Classroom discourse: The language of teaching and learning.* Portsmouth, NH: Heinemann.

Clarkson, P. (1991). *Bilingualism and mathematics learning.* Geelong, Victoria: Deakin University Press.

Clarkson, P., & Galbraith, P. (1992). Bilingualism and mathematics learning: Another perspective. *Journal for Research in Mathematics Education, 23*(1), 34–44.

Crowhurst, M. (1994). *Language and learning across the curriculum.* Scarborough, Ontario: Allyn and Bacon.

Cummins, J. (1976). The influence of bilingualism on cognitive growth. *Working Papers on Bilingualism, 9,* 1–43.

Cummins, J. (1979). Cognitive/academic language proficiency, linguistic interdependence, the optimum age question, and some other matters. *Working Papers on Bilingualism, 19,* 197–205.

Cummins, J. (1994). Semilingualism. In R. Asher (Ed.), *International encyclopedia of language and linguistics* (2nd ed.) (pp. 3812–3814). Oxford: Elsevier Science Ltd.

Cummins, J. (2000). *Language, power, and pedagogy.* Buffalo, NY: Multilingual Matters.

D'Ambrosio, U. (1991). Ethnomathematics and its place in the history and pedagogy of mathematics. In M. Harris (Ed.), *Schools, mathematics and work* (pp. 15–25). Bristol, PA: Falmer Press.

Dawe, L. (1983). Bilingualism and mathematical reasoning in English as a second language. *Educational Studies in Mathematics, 14*(4), 325–353.

De Avila, E., & Duncan, S. (1981a). Bilingualism and the metaset. In R. Durán (Ed.), *Latino language and communicative behavior* (pp. 337–354). Norwood, NJ: Ablex Publishing Corporation.

De Avila, E., & Duncan, S. (1981b). *A convergent approach to oral language assessment: Theoretical and technical specification on the Language Assessment Scales (LAS)* Form A (Stock 621). San Rafael, CA: Linguametrics Group.

Duncan, S., & De Avila, E. (1986). *Pre-LAS user's manual* (Form A). San Rafael, CA: Linguametrics Group.

Duncan, S., & De Avila, E. (1987). *Pre-LAS Español user's manual* (Form A). San Rafael, CA: Linguametrics Group.

Erickson, F. (1986). Qualitative methods in research on teaching. In M. Wittrock (Ed.), *Handbook of research on teaching* (3rd ed.) (pp. 119–161). New York: MacMillan Publishing Company.

Ellerton, N., & Clements, M. (1991). *Mathematics in language: A review of language factors in mathematics learning.* Geelong, Victoria: Deakin University.

Finegan, E., & Besnier, N. (1989). *Language: Its structure and use.* New York: Harcourt Brace Jovanovich.

Garcia, E., & Gonzalez, R. (1995). Issues in systemic reform for culturally and linguistically diverse students. *Teachers College Record, 96*(3), 418–431.

Gee, J. (1996). *Social linguistics and literacies: Ideology in discourses* (3rd ed.). London: The Falmer Press.

González, N. (1995). Processual approaches to multicultural education. *Journal of Applied Behavioral Science, 31*(2), 234–244.

González, N., Andrade, R., Civil, M., & Moll, L.C. (2001). Bridging funds of distributed knowledge: Creating zones of practices in mathematics. *Journal of Education for Students Placed at Risk, 6,* 115–132.

Grosjean, F. (1999). Individual bilingualism. In B. Spolsky (Ed.), *Concise encyclopedia of educational linguistics* (pp. 284–290). London: Elsevier.

Gumperz, J. (1982). *Discourse strategies.* New York: Cambridge University Press.

Gutiérrez, K., Baquedano-Lopez, P., & Alvarez, H. (2001). Literacy as hybridity: Moving beyond bilingualism in urban classrooms. In M. de la Luz Reyes & J. Halcon (Eds.), *The best for our children: Critical perspectives on literacy for Latino students* (pp. 122–141). New York: Teachers College Press.

Gutiérrez, K., & Rogoff, B. (2003). Cultural ways of learning: Individual traits or repertoires of practice? *Educational Researcher, 32*(5), 19–25.

Hakuta, K., & Cancino, H. (1977). Trends in second-language-acquisition research. *Harvard Educational Review, 47*(3), 294–316.

Hakuta, K., & McLaughlin, B. (1996). Bilingualism and second language learning: Seven tensions that define research. In D. Berliner & R. C. Calfe (Eds.), *Handbook of Educational Psychology* (pp. 603–621). New York: Macmillan.

Hall, R. (2000). Video recording as theory. In R. Lesh & A. Kelly (Eds.), *Handbook of research design in mathematics and science education* (pp. 647–664). New Jersey: Lawrence Erlbaum Associates, Inc.

Halliday, M. A. K. (1978). Sociolinguistics aspects of mathematical education. In M. Halliday (Ed.), *The social interpretation of language and meaning* (pp. 194–204). London: University Park Press.

Heath, S. B. (1983). *Ways with words.* Cambridge: Cambridge University Press.

Jones, P. (1982). Learning mathematics in a second language: A problem with more and less. *Educational Studies in Mathematics, 13,* 269–288.

Khisty, L. (1995). Making inequality: Issues of language and meanings in mathematics teaching with Hispanic students. In W. G. Secada, E. Fennema, & L. B. Adajian (Eds.), *New directions for equity in mathematics education* (pp. 279–297). New York: Cambridge University Press.

Khisty, L. (2001). Effective teachers of second language learners in mathematics. In *Proceedings of the 25th Conference of the International Group for the Psychology of Mathematics Education* (pp. 225–232). Utrecht, The Netherlands: The Freudenthal Institute, Utrecht University.

Khisty, L., & Chval, K. (2002). Pedagogic discourse and equity in mathematics: When teachers' talk matters. *Mathematics Education Research Journal, 14*(3), 154–168.

Lee, C. (1993). *Signifying as a scaffold for literary interpretation: The pedagogical implications of an African American discourse genre.* (Research Rep. No. 26). Urbana, IL: National Council of Teachers of English.

MacSwan, J. (2000). The threshold hypothesis, semilingualism, and other contributions to a deficit view of linguistic minorities. *Hispanic Journal of Behavioral Sciences, 22*(91), 3–45.

McDermott, R., Gospodinoff, K., & Aron, J. (1978). Criteria for an ethnographically adequate description of concerted activities and their contexts. *Semiotica, 24,* 245–275.

McLain, L., & Huang, J. (1982). Speed of simple arithmetic in bilinguals. *Memory and Cognition, 10,* 591–596.

Marsh, L., & Maki, R. (1976). Efficiency of arithmetic operations in bilinguals as a function of language. *Memory and Cognition, 4,* 459–464.

Mehan, H. (1979). *Learning lessons: Social organization in the classroom.* Cambridge, MA: Harvard University Press.

Moll, L., Amanti, C., Neff, D., & González, N. (1992). Funds of knowledge for teaching: Using a qualitative approach to connect homes and classrooms. *Theory into Practice, 31,* 132–141.

Moschkovich, J. N. (1996). Moving up and getting steeper: Negotiating shared descriptions of linear graphs. *The Journal of the Learning Sciences, 5*(3), 239–277.

Moschkovich, J. N. (1999). Supporting the participation of English language learners in mathematical discussions. *For the Learning of Mathematics, 19*(1), 11–19.

Moschkovich, J. N. (2000). Learning mathematics in two languages: Moving from obstacles to resources. In W. Secada (Ed.), *Changing faces of mathematics (Vol. 1): Perspectives on multiculturalism and gender equity* (pp. 85–93). Reston, VA: NCTM.

Moschkovich, J. N. (2002). A situated and sociocultural perspective on bilingual mathematics learners. *Mathematical Thinking and Learning,* Special issue on Diversity, Equity, and Mathematical Learning, N. Nasir & P. Cobb (Eds.), *4*(2&3), 189–212.

Moschkovich, J. N. (2004). Appropriating mathematical practices: A case study of learning to use and explore functions through interaction with a tutor. *Educational Studies in Mathematics, 5,* 49–80.

Moschkovich, J. N. (2007a). Bilingual mathematics learners: How views of language, bilingual learners, and mathematical communication impact instruction. In N. Nasir & P. Cobb (Eds.), *Diversity, equity, and access to mathematical ideas* (pp. 89–104). New York: Teachers College Press.

Moschkovich, J. N. (2007b). Examining mathematical Discourse practices. *For the Learning of Mathematics, 27*(1), 24–30.

Moschkovich, J. N. (2007c). Using two languages while learning mathematics, *Educational Studies in Mathematics, 64*(2), 121–144.

Moschkovich, J. N. (2008). "I went by twos, he went by one:" Multiple interpretations of inscriptions as resources for mathematical discussions. *The Journal of the Learning Sciences, 17*(4), 551–587.

Moschkovich, J. N., & Brenner, M. (2000). Integrating a naturalistic paradigm into research on mathematics and science cognition and learning. In R. Lesh & A. Kelly (Eds.). *Handbook of Research Design in Mathematics & Science Education* (pp. 457–486). Mahwah, NJ: Lawrence Erlbaum Associates, Inc.

National Center for Educational Statistics. (1995). *The condition of education 1995.* Washington, DC: U.S. Department of Education, OERI.

Nunes, T., Schliemann, A., & Carraher, D. (1993). *Street mathematics and school mathematics.* Cambridge: Cambridge University Press.

Ochs, E. (1979). Transcription as theory. In E. Ochs & B. Schieffelin (Eds.), *Developmental pragmatics* (pp. 41–72). New York: Academic Press.

O'Connor, M. C. (1999). Language socialization in the mathematics classroom. Discourse practices and mathematical thinking. In M. Lampert & M. Blunk (Eds.), *Talking mathematics* (pp. 17–55). New York: Cambridge University Press.

O'Halloran, K. (1999). Towards a systemic functional analysis of multisemiotic mathematics texts. *Semiotica, 124*(1/2), 1–29.

Orey, D. (2003). *The algorithm collection project: An exploration of the ethnomathematics of basic number sense across cultures.* Paper presented at CERME 3: Third Conference of the European Society for Research in Mathematics Education, Bellaria, Italy.

Paulston, C. (1982). *Swedish research and debate about bilingualism.* Skoloverstyrselsen: National Swedish Board of Education.

Pew Hispanic Center/Kaiser Family Foundation. (2004). National survey of Latinos: Education. Washington, DC: Pew Hispanic Center. Available at http://pewhispanic.org/files/reports/25.pdf. Retrieved on August 12, 2009.

Pimm, D. (1987). *Speaking mathematically: Communication in mathematics classrooms.* London: Routledge.

Poland, B. (2002). Transcription quality. In J. Gubrium & J. Hosltein (Eds.), *Handbook of interview research context and method* (pp. 629–649). Thousand Oaks, CA: Sage.

Pressley, M. (2000). What should comprehension instruction be the instruction of? In M. Kamil, P. Mosenthal, P. D. Pearson, & R. Barr (Eds.), *Handbook of reading research* (Vol. III, pp. 545–561). Mahwah, NJ: Lawrence Erlbaum Associates.

Roberts, T. (1998). Mathematical registers in aboriginal languages. *For the Learning of Mathematics, 18*(1), 10–16.

Sanchez, R. (1994). *Chicano discourse: Socio-historic perspectives.* Houston, TX: Arte Público Press.

Savignon, S. (1991). Communicative language teaching: State of the art. *TESOL Quarterly, 25*(2), 261–277.

Scarcella, R. (2003). *Academic English: A conceptual framework.* Technical Report 2003-1, University of California Linguistic Minority Research Institute. Irvine, CA.

Schleppegrell, M. (2007). The linguistic challenges of mathematics teaching and learning: A research review. *Reading & Writing Quarterly, 23,* 139–159.

Scribner, S. (1984). Studying working intelligence. In B. Rogoff & J. Lave (Eds.), *Everyday cognition: Its development in social context* (pp. 9–40). Cambridge: Harvard University Press.

Secada, W. (1983). *The educational background of limited English proficient students: Implications for the arithmetic classroom.* Arlington Heights, IL: Bilingual Education Service Center. Washington, DC: Office of Bilingual Education and Minority Languages Affairs.

Secada, W. (1991). Degree of bilingualism and arithmetic problem solving in Hispanic first graders. *Elementary School Journal, 92*(2), 213–231.

Setati, M. (1998). Code-switching and mathematical meaning in a senior primary class of second language learners. *For the Learning of Mathematics, 18*(1), 34–40.

Setati, M., & Adler, J. (2001). Between languages and discourses: Code switching practices in primary classrooms in South Africa. *Educational Studies in Mathematics, 43,* 243–269.

Skutnabb-Kangas, T. (1984). *Bilingualism or not: The education of minorities.* Clevedon: Multilingual Matters.

Souviney, R. (1983). Mathematics achievement, language and cognitive development: Classroom practices in Papua New Guinea. *Educational Studies in Mathematics, 14*(2), 183–212.

Spanos, G., & Crandall, J. (1990). Language and problem solving: Some examples from math and science. In A. M. Padilla, H. H. Fairchild, & C. M. Valadez (Eds.), *Bilingual education: Issues and strategies* (pp. 157–170). Beverly Hill, CA: Sage.

Spanos, G., Rhodes, N. C., Dale, T. C., & Crandall, J. (1988). Linguistic features of mathematical problem solving: Insights and applications. In R. Cocking & J. Mestre (Eds.), *Linguistic and cultural influences on learning mathematics* (pp. 221–240). Hillsdale, NJ: Lawrence Erlbaum.

Swain, M. (2001). Integrating language and content teaching through collaborative tasks. *Canadian Modern Language Review, 58*(1), 44–63.

Torres, L. (1997). *Puerto Rican discourse: A sociolinguistic study of a New York suburb.* Mahwah, NJ: Lawrence Erlbaum.

Ulewicz, M., & Beatty, A. (Eds.) (2001). *The power of video in international comparative research in education.* Washington DC: National Academy Press.

Valdés-Fallis, G. (1978). Code switching and the classroom teacher. *Language in education: Theory and practice* (Vol. 4). Wellington, VA: Center for Applied Linguistics. (ERIC Document Reproduction Service No. ED153506)

Valdés-Fallis, G. (1979). Social interaction and code switching patterns: A case study of Spanish/English alternation. In G. D. Keller, R.V. Teichner, & S. Viera (Eds.), *Bilingualism in the bicentennial and beyond* (pp. 86–96). Jamaica, NY: Bilingualism Press.

Vogt, L., Jordan, C., & Tharp, R. (1987). Explaining school failure, producing school success: Two cases. *Anthropology & Education Quarterly, 18*(4), 276–286.

Zentella, A. (1981). Tá bien, you could answer me en cualquier idioma: Puerto Rican code switching in bilingual classrooms. In R. Durán (Ed.), *Latino language and communicative behavior* (pp. 109–130). Norwood, NJ: Ablex Publishing Corporation.

Zentella, A. C. (1997). *Growing up bilingual: Puerto Rican children in New York.* Malden, MA: Blackwell Publishers.

CHAPTER 2

DEVELOPING A MATHEMATICAL VISION

Mathematics as a Discursive and Embodied Practice

**Kris D. Gutiérrez, Tesha Sengupta-Irving,
and Jack Dieckmann**

ABSTRACT

This chapter examines major lines of inquiry in mathematics education through the prism of cultural historical activity theory, focusing on the language and discursive practices in the teaching and learning of school mathematics. We make an analytic distinction between the language *in* and *of* mathematics learning in classrooms, noting the pitfalls of dichotomizing the language *of* the classroom and the language *in* mathematical learning or ignoring their interrelations. Specifically, we reviewed work that framed the role of everyday discourse practices as supporting the development of scientific discourse practices. In line with several mathematics education scholars, we challenge this framing by revisiting the theoretical principles from the work of Vygotsky, Engeström, and Cole, among others, to show how scientific or school-based mathematical learning "grows down into" the everyday, and thus

Language and Mathematics Education, pages 29–71

argue their relation. We proffer the notion and importance of a *mathematical vision* for mathematics learning that is robust, multi-semiotic, and embodied. We ground this review in the assumption that developing a mathematical vision (i.e., knowing the practice of mathematics) involves a socially organized way of seeing, understanding, envisioning, and doing mathematics in ways that are accountable to the distinct norms of the mathematical community.

> *A real concept is an image of an objective thing in its complexity.*
> *Only when we recognize the thing in all its connection and relation,*
> *only when this diversity is synthesized in a word, in an integral image*
> *through the multitude of determinations, do we develop a concept.*
> —Vygotsky (1997, p. 53)

INTRODUCTION

In this chapter, we address those areas of scholarship in the study of language and mathematical discourse that we believe are opening up new lines of inquiry. These topics include the dilemma of working with students' everyday language and discourses while helping them develop competence in the language and discourse of mathematics, work that examines mathematics as an "embodied" practice, as well as work that takes an approach to mathematics as a multimodal and multi-semiotic activity. We were initially motivated by a general interest in the role of communication practices in learning activity and by an increased interest in how action, cognition, and activity are socioculturally organized in mathematics-related thinking and learning activity. An initial review was conducted[1] to learn more about what existing research had to say about the role of language, academic and mathematical literacy, and discourses in mathematical learning. This preliminary review generated a keen interest in learning more about new lines of inquiry in mathematics education but also raised a number of questions and observations about current understandings of language and mathematical discourse that we believed warranted further examination. For example, we wondered why mathematical discourse is defined and examined in such diverse ways in the mathematics education literature. We also noted a blurring of the distinction between classroom discourse and mathematical discourse practices. We observed that mathematics learning as an embodied activity, as well as how language, discourse, and other tools work together to mediate mathematics learning, were largely understudied.

As a result of this review and our own theoretical orientation, we began the next phase of the review with a slightly more expansive scope. We were interested in work that conceived of communication in human learning activity more broadly—as Roth (2001) argues, work that moves us beyond "categorical features of language to include what are held by some to be the embodied expression of human experience" (p. 368). In this chapter, the notion of embodied learning spans several disciplines and includes, for example, cognitive psychology (Nuñez, Edwards, & Matos, 1999; Schwartz & Black, 1999) and ethnomethodology (Livingston, 1999, 2006), which present a theory of embodied cognition or embodied mathematics, and Goodwin's (2000, 2006) linguistic anthropological approach to embodiment, which is a more holistic framework for understanding human accomplishment at work. From these perspectives, the accomplishment of human cognitive activity occurs across distributed systems that include joint activity, language, behaviors, tools, and the social organization of action among participants (Goodwin, 2000).

Our thinking about how language, discourse, and embodied practices influence mathematics learning evolved in the course of reading relevant scholarly work that identified these key lines of inquiry as potentially fruitful. To conduct our review, we first interviewed scholars whose work and recommendations we believed would guide our assessment of the relevant literature.[2] We read widely and synthesized what we learned using a cultural historical activity theoretical perspective to better understand how the various authors made sense of mathematics as an activity system (Cole & Engeström, 1993; Engeström, 1987, 2005; Leont'ev, 1978). Using activity theory as an analytic heuristic also helped us account for the multiple dimensions that constitute mathematics learning activity (e.g., subjects/participants, mediating artifacts, rules, community, division of labor, and object of the activity)—dimensions addressed variably by mathematics education researchers.

Guided by these heuristic tools, we call attention to promising understandings of language and learning in the context of mathematics education and argue for their further examination. Further, our review of the literature identified issues that we believe are especially relevant in working with non-dominant student populations and within the context of constraining educational policies and practices. We briefly summarize some of these issues.

First, the issue of how scientific or school-based concepts/ideas develop from everyday concepts, as well as their relation, remains an enduring empirical question. What are the affordances of understanding everyday and mathematical discourses relationally? And if, in fact, scientific (literate) discourse develops from the everyday (colloquial) discourse, as Sfard and Cole (2003) put it, what are the implications for students who bring varying repertoires of practice to school? We believe this remains an exceedingly important question in mathematics education, particularly in light of the dearth of studies of mathematical learning in non-dominant U.S. communities.

In a related point, the issue of what transfers or what mathematical knowledge and discourses take hold across practices and settings remains a persistent empirical question.[3] While the idea of everyday and school-based discourses and practices is taken up in the literature we reviewed, the issue of what is portable across everyday and school tasks and contexts or how to make what takes hold visible to others remains largely a black box. In other words, what is less understood is what carries and how it happens. From our perspective, the focus should be on what takes hold across tasks and practices, as a way to capture students' repertoires of practice.

Second, if mathematical understanding involves multiple modalities and artifacts, including oral and written language, gesture, the body, symbols, representations, and the like, then studies that take an interdisciplinary approach would enhance the field of mathematics education research. In particular, situated cognition, and anthropological, activity theoretic, social practice, and ethnomathematical perspectives could deepen our understanding of mathematics as a multimodal practice. This view of mathematics as a multimodal and multi-semiotic activity, we believe, moves us toward a more expansive way of approaching mathematical communication and learning.

Taken together, we believe that the above topics constitute important areas of study in mathematics education, especially in relation to issues of equity. Creating robust and equitable contexts for learning mathematics for all students is an increasingly important issue in mathematics education. Thus, understanding how to create and sustain environments for learning mathematics, as well as how the social organization of these environments facilitates or interferes with cognitive work, should be of central concern in mathematics education research and practice. In the same vein, we would want to know more about what forms of mediation support deep mathematics learning that leads to transformational understandings of concepts and practices. How are people made "smart" by the use of artifacts and participation in particular social groups and settings? In particular, what role do language and mathematical discourse play in mathematics learning processes?

Our questions and reading of the literature are informed by our own theoretical and disciplinary knowledge and, as such, this chapter reflects those orientations. We take a sociocultural and activity theoretic approach to learning and development in the domains of literacy and mathematics. Equity is an organizing principle across our review. Although we brought our collective set of theoretical and methodological lenses to this discussion, we agreed that our contribution would focus on how viewing mathematics as culturally and historically structured would be useful in understanding mathematical practices and the resources that mediate learning therein. In line with this view, we examine how the current literature addresses the ways language and other tools function to support mathematical understanding.

It is important to clarify that this chapter was not intended to be an exhaustive review of research in mathematical learning, as there are a number of important and comprehensive reviews of this literature (Carpenter, Fennema, & Franke, 1996; Confrey, 1990; Durkin & Shire, 1991; Franke, Kazemi, & Battey, 2007; Hufferd-Ackles, Fuson, & Sherin, 2004; Schoenfeld, 1992). The approach we take is different than is typically found in such reviews. This chapter is organized in two main sections. The first part describes the orienting framework that guided our attention to language, discourse, and embodied practices as mediating mathematics learning. This discussion illustrates a growing understanding of mathematics as a cultural activity, a result of the turn toward sociocultural views of mathematics learning. The second portion of the chapter highlights work that contributed to an understanding of the tools that mediate mathematics learning. The literature presented here elaborates three key ideas relevant to a cultural historical view of mathematics: (1) mathematical discourse as a cultural activity (i.e., talking mathematics is learning mathematics); (2) everyday and scientific mathematical discourse practices; and (3) mathematics as a multimodal activity (encompassing both embodiment and multi-semiotics).

ORIENTING FRAMEWORK

In this review, we advance work that views mathematics as a multimodal and multi-semiotic activity. To do so, we take a cultural historical activity theoretic (CHAT) approach to learning in which the structure and development of human cognition (thinking, learning, and acting) emerges through culturally mediated, historically developing, practical activity (Cole, 1996). The use of signs and tools is fundamental to the concept of a mediated human environment (Vygotsky, 1978).

> By "signs" he [Vygotsky] referred to socially created symbol systems such as language, writing, and number systems, which emerge over the course of history and vary from one society to another. Mental processes always involve signs, just as action on the environment always involves physical instruments (if only a human hand). (Scribner & Cole, 1981, p. 8)

This view of human development attempts to resolve the mind and society divide perpetuated by accounts of cultural and psychological change (Scribner & Cole, 1981; Vygotsky, 1962, 1978). Of significance to this chapter, language, the tool of all mediational tools (Cole & Engeström, 1993), assumes a critical role in the mediation of everyday and school life and plays a crucial role in the development of higher psychological processes. However, analysis of human activity often privileges language over other semiotic means, rendering it autonomous, where everything that isn't language is lumped into the category of context (Goodwin, 2000).

In line with this view and the interdisciplinary character of a cultural historical approach, we argue against viewing language autonomously (Scribner & Cole, 1981) from other semiotic means. For example, Kieran, Forman, & Sfard (2002) and, in particular, Sfard (2007) use a discursive psychological approach inspired by Vygotsky to make a useful distinction—one not uniformly shared in mathematics education literature—language is a tool, whereas discourse is an activity in which the tool is used or mediates. Since students are socialized both to and through (Ochs, 1988) mathematical discourse practices, this perspective importantly recognizes the role language practices play in the overall development of mathematical discourse.

We believe that approaches that look at language practices and discourse *and* that embrace the complex linguistic nature of mathematical activity will be particularly productive in discussions of mathematics learning. At the same time, we argue against analytic and disciplinary views that do not take into account the culturally-specific ways participants simultaneously use multiple semiotic resources in activity. In short, our interest in artifacts and their mediating role in mathematical learning beg for an accounting of mathematics as a mediated cultural activity, as discussed briefly below.

THE CULTURAL DIMENSIONS OF MATHEMATICS

In this brief review, we note work that addressed the cultural nature of mathematics and the influence of culture on mathematical learning. We include classical pieces such as Saxe's (1988, 1991) work on Brazilian candy sellers, Scribner's (1983/1997) work on American dairy farmers, Cole's early work on quantitative abilities in Liberia (Gay & Cole, 1967), Lave's (1988) work on the mathematics of supermarket shopping, Nasir's (2000, 2002; Nasir & Hand, 2004) work on mathematical understandings and practices of African American basketball and dominoes players, and Goldman's (2005) work on families' mathematical practices as examples of a cultural and social practice view of mathematics. One common theme across this work is that mathematical thinking and learning are deeply tied to the activities of which participants are a part.

Based on decades of empirical work on the role of culture in cognitive development, Cole and his colleagues argued that understanding the role of culture in the development of mathematical skills requires using the everyday tasks that are already part of people's repertoires of practice (Gutiérrez & Rogoff, 2003) as a starting point (Cole, Gay, Glick, & Sharp, 1971). Taken together, this body of work emphasizes the salience of understanding mathematical learning in the contexts of its practice, locally and historically. Moreover, such work moves our focus away from autonomous models of mathematics to viewing mathematical thinking in relation to the larger social organization of the activities within which it is used.

Conceptualizing mathematics as a cultural system has been addressed interdisciplinarily. D'Ambrosio (1984), who introduced the idea of ethnomathematics in the 1980s, was interested in how cultural groups vary in their use of mathematics: that is, how cultural groups "mathematize"— count, measure, relate, classify, and infer. For D'Ambrosio,

> ... mathematical practices differ from one cultural group to another. At this level, Mathematics comes close to being a variant of common language associated with the concept of codifying popular practices and daily needs. These would include uses of numbers, of quantities, the capability of qualifying and quantifying and some patterns of inference. This is what we call, borrowing a term introduced by Tadasu Kawaguchi, matheracy. (p. 44)

From D'Ambrosio's perspective, social groups and age groups (e.g., children, farmers, engineers, and professional groups of people) enact a culture with their own specific jargon, patterns of doing, symbols and code, and their own way of mathematizing. Children, then, come to school with a mathematical culture of their own (D'Ambrosio, 1984). The interest in the cultural dimensions of mathematics in communities helped to make the most of the contrasts between home and school mathematics rather than viewing them as deficits.

THE SOCIAL TURN IN MATHEMATICS

The social turn in mathematics education refocused attention on the social organization of cognition, interaction, and embodied activity and has looked to interdisciplinary perspectives, particularly sociological, anthropological, and cultural approaches to the teaching and learning of mathematics, for more robust understandings of how students learn school-based mathematics. Brown, Stein, and Forman's (1996) study, for example, employed a sociocultural perspective to document change in one middle-school mathematics reform project. By examining the social organization of the classroom, they identified forms of assistance that mediated mathematics teaching and learning in these classrooms. (See Lerman, 1996, 2002, for a more comprehensive review of this issue).

With the social turn in mathematics, we find interesting hybrids of cognitive and social perspectives. Cobb, Wood and Yackel (1993), for example, elaborated their view that mathematics is a social activity or community project, in addition to being an individually constructed activity. Informed by a cognitive constructivist perspective, the authors conducted a two-year study of a second grade mathematics classroom characterized by the absence of grading and individual pencil-and-paper work to study mathematics as a socially accomplished activity. Through an analysis of classroom talk, Cobb and his colleagues emphasize that "the reflexive relation between the individual and the social holds for students' mathematical development, as well as for the teachers' pedagogical development" (1993, p. 96).

One of the contributions of this study was its analysis of the classroom participants' interactional patterns around two major linguistic activities: "talking about math," and "talking about talking about math." Whereas the former activity often involved open-ended questioning and encouraged inquiry, the latter followed a traditional IRE (Initiation, Response, Evaluation) pattern of discourse (Mehan, 1978, 1979). Of the many claims presented to affirm mathematics as a social activity, the authors focus specifically on the development of social norms constructed in small group and whole class instruction (e.g., we work to find our answers, we do not just take them from others; mistakes are welcome and necessary, etc.). In this study the classroom is conceptualized as a community that is unique in its own right such that young people are not viewed as mini-mathematicians engaging in practices that approximate mathematicians' behaviors. Framed as the development of community and shared mathematical practices, the authors assume a socio-cognitive perspective in which they link their views with a sociocultural approach to learning: "Such a view is, as we have seen, compatible with Vygotsky's analysis of the social situation of development" (Cob et al., 1993, p. 107).

O'Connor (1998) presents a review of the approaches taken in exploring the connection between communication about mathematics and learning mathematics in classrooms. She suggests that research from the process–product tradition tends to demonstrate that students who elaborate explanations show modest achievement gains, though the author does not explain why. However, more recent research traditions, including those based on sociocultural theory, suggest the study of individual learning leads to a study of the social: "When an individual has been effectively 'inserted' into a recognized social practice, the individual will grow to participate in the embedded forms of reasoning and behaving, as these are called for by the social practice in question" (O'Connor, 1988, pp. 23–24). O'Connor then takes up this turn toward the social through an examination of classroom talk as a means of deepening students' mathematics' learning. In so doing, she questions how *protoforms* of mathematical discourse, mathematical practices that are more readily understood in an everyday sense, and familiarity with such protoforms may translate into the habits of mind required in mathematical thinking. In particular, O'Connor questions how protoforms of discipline-based argumentation and negotiated defining transfer to mathematical practices in the classroom.

Such work has inspired a shift from thinking of mathematical thinking and learning as organized in the mental life of individuals to understanding these as social and discursive accomplishments, which are socially mediated, and in which talk, the body, diagrams, representations, objects and the social situation of development play an important role. From this expanded view, mathematics is a multimodal system that requires various modes of communication. (See the work of Adler, 1998; Forman, 1996; Lerman, 1996, 2002; Morgan, 1996, 1998; Nasir, 2000, 2002; O'Halloran, 2005; Saxe, 1988, 1991; Schmittau, 1993, 2004; Sfard, 2000; and Street, Rogers, & Baker, 2006 for examples of relevant work that share aspects of this social and cultural approach to mathematical learning).

One of the most prevalent ways researchers responded to the social turn in mathematics education was to focus on talk and interaction in mathematics classrooms. Researchers such as Ball (1993) elaborated the relationship between discourse, content, and community across her research to illustrate how these elements work in concert to help students develop what Schoenfeld (1992) calls a "mathematical point of view" (p. 335). Developing this stance involves valuing the process of mathematization, abstraction, as well as a predilection to apply them. In her work, Ball characterizes mathematics as a socially constructed process in which sense-making is both individual and consensual. From this perspective, community is crucial to learning mathematics insofar as the kind of classrooms Ball proposes positions mathematical reasoning as the standard for validity rather than teacher authority. Here, classroom mathematical discourse is essentially a process of establishing truth claims about mathematical situations. In particular, Ball emphasizes the complex listening required of teachers using classroom discourse to guide instruction. Teachers must attend carefully to what students are saying and meaning with an eye toward "the mathematical horizon." In this work, Ball also identifies a contradiction taken up by a number of mathematics education researchers: how to integrate students' inventiveness in mathematics with the conventions of the wider mathematics community.

Examining Language *in* and *of* Mathematics Classrooms

In general, the literature addressing the social nature of mathematical learning has focused on language *in* mathematics learning activity—that is, in practices such as argumentation, proof and justification, generalization, negotiated defining, representation, and so on that constitute mathematical thinking. Alternatively, this literature focused on language *of* the mathematics classroom, that is, the ways in which language and discourse are regulated by teachers and students. We think this distinction, advanced by Hymes (1977), Green and Dixon (1994), and Lin (1994), is useful in examining work on the development of mathematical discourse (language *in*) from the more general classroom discourse practices (language *of*).

Pimm's (1987, 1991) work proved critical to examinations of mathematics and language. He is most notably credited in the field of mathematics education for exploring the familiar claim that mathematics is a language. Through his work, Pimm suggests mathematics is best understood as a register that carries "a set of meanings that is appropriate to a particular *function* of language" (1987, p. 17). Registers have to do with the social usage of particular words and expressions, ways of talking but also ways of meaning. This register is distinguished by way of word meaning and, most importantly, symbol use (Pimm, 1987). By asking what it means to become fluent in this register, Pimm's attention turns to the language and practices of mathematics classrooms. He concludes, "Pupils learning mathematics in school in part are attempting to acquire communicative competence in the mathematical register, and classroom activities can be carefully examined from this perspective in order to see what opportunities they are offering pupils for learning" (1991, p. 22). Indeed, this call to focus on the language *of* and *in* mathematics classrooms was subsequently taken up by others in the field.

In her work, Lampert (1998) draws on her elementary mathematics teaching to make visible the moment-to-moment construction of mathematical talk, including the inherent complexity, contradictions, and dilemmas of teaching. Lampert illustrates aspects of mathematical talk that include position taking, question asking, proof and justification, expanding ideas, use of evidence, assertions, conjectures, symbolic reference, and so on. By emphasizing the need for classroom-based research, Lampert argues for attention to the influence of language and discourse on mathematical learning from the teacher's point of view. Invoking the work of Luria, Vygotsky, and Bakhtin, Lampert suggests that "conversation and culture have become inseparable foci for investigation in classroom research" (p. 8) to the extent that the very definition of mathematics learning now puts discourse at its center. For Lampert, the importance of attending to discourse was inspired by the need to account for the social and the cultural, a renewed interest by researchers on talk and interaction, as well as reform-based recommendations for changes in U.S. mathematics education (National Council of Teachers of Mathematics [NCTM], 1991, 2000).

Kerslake's (1991) focus on the language of learning fractions is representative of work that looks at language for a particular mathematical topic to articulate its features. Such efforts to examine the specific language of a topic help to identify how language and use of other tools become sites for understanding student misconceptions. Kerslake focuses on the world of fractions and narrows the scope of inquiry to the specific words, phrases, and objects students use in studying them. Based on a series of student interviews, the author suggests that students stumble in conceiving of fractions as numbers because they perceive them as "broken numbers." Interestingly, Kerslake also reveals how students tend to rely on the everyday language of "sharing" to describe division and surmises this happens because sharing is likely to have been students' first experience of dividing. Kerslake concludes by taking aim at classroom algorithmic practices, saying they clutter students' speech with unnecessary jargon and obscure mathematical ideas. Instead, Kerslake calls for a closer look at how students think of and talk about fractions through the course of their learning.

Ferrari (2004) similarly focuses on language *in* mathematics learning activity as a way of understanding why students struggle in advanced mathematics. This focus led Ferrari to an articulation of how students' discourse practices matter in their ability to master the discipline. Drawing upon a theoretical frame of functional linguistics, Ferrari suggests it is students' reliance on interpersonal communicative competence over fluency in a more organized, logical, specific mathematical register that accounts for their failure. He suggests that a more customary explanation for student failure in advanced mathematics is that students fail to master the content in their high school mathematics curriculum. Instead, he argues, it is more productive to consider the need for explicitly teaching students how to translate between everyday and mathematics registers. Ferrari concludes by noting that educators should take seriously the need for students to become familiar with the literate register of mathematics as an object of explicit learning, rather than mistaking fluency in the mathematical register as a natural byproduct of engaging in the discipline. In this way, he suggests that a more robust understanding of student achievement and failure in mathematics occurs through considering students' competence in the specific language and discourse practices of the domain.

McNair (1998) similarly focuses on language and discourse as the site for evaluating students' learning of mathematics. McNair suggests that the central enterprise of learning mathematics is having students participate in the discourse practices of the discipline. By this rubric, knowing mathematics means participating in cognitive and communicative processes driven by a mathematics frame, which is characterized by a valuing of objectivity of number, logical ordering of the universe, precision, definition, dichotomy, and abstract objectivity. McNair examines the creation of text, verbal portions of discussions that can be recorded and analyzed (i.e., utterances), to better understand the different types of discourse requirements imposed by different mathematics problems and how these effect children's inclination to create their own mathematics contexts. If, as McNair argues, the goal of mathematical communication is to create an abstract context in which arguments can be constructed and resolved, then learning to communicate mathematically depends on students' and teachers' abilities to work within various mathematics contexts. Inspired by Goffman's (1974) notion of frames, McNair considers multiple frames in which students operate while engaging mathematically: the *peer* frame, the *calculational* frame, the *problem* frame, and so on. Though all are important, the mathematics frame is characterized by a sustained focus on specific mathematical relationships inscribed within a problem or task. According to McNair, it is the ability to work within a mathematical frame, to create and manipulate mathematics contexts and produce text, that ultimately marks a successful mathematics learner.

Kieren (1999) similarly focuses on the language *in* mathematics classrooms. Yet Kieren's work makes more prominent how students' language and discourse practices are influenced by their membership in a classroom community. Indeed, for Kieren, mathematical learning involves the co-emergence of language practices shaped by both the individual and the environment. Previously, inscriptions were viewed as relating solely to cognition. Kieren provides illustrative examples of how inscriptions serve as a living record of learning (in this case, learning fractions). In this study, some students used language and other representations to record their classroom experiences; others used fractional language that seemed set apart from classroom experiences; and still others, using similar language, seemed not to reason with fractions at all; thus, as Kieren suggests, they did not *bring forth a world of mathematics*. Kieren asserts that when looking at students' fractional language use in various settings, a researcher or teacher needs to be aware that the students perceive those settings in their own terms, and their fractional language reflects the nature of that setting, their lived experience in it, and the students' own structures. Here, Kieren encourages us to look beyond the mathematical symbol or its referent to the person using the language in mathematically knowing ways.

Burton (2001) looks at both the language *in* and *of* mathematics classrooms in her articulation of mathematical narrative and the actions of agentic mathematical learners. Burton asserts that *narrative* is what we all use to impose coherent meaning on experiences, including mathematics.

She adopts the perspective, shared by Wenger (1998), that a dual process of participation and reification constitutes the negotiation of meaning. "In this way, mathematics becomes a socio-cultural story told in a socially negotiable context" (Burton, 2001, p. 4). Burton suggests that the key features of discourse that make a narrative *mathematical* include: presupposition (creation of implicit meanings), subjectification (centralizing the practices of inquiry), and multiple perspective (recognition and celebration of alternate views). In her analysis of classroom learning and interaction, Burton shows the predominant actions that agentic children do in mathematics are: (1) authoring, (2) sense-making, (3) collaborating, and (4) using nonverbal narratives. However, she notes, typical classroom practices seem to thwart the use of students' narratives and, thus, students as meaning-makers and mathematical storytellers.

The increased interest in the social dimensions of mathematical thinking and learning, the blurring or bringing together of the language *in* and *of* the classroom, is exemplified in the work of Yackel and Cobb (1996), who developed the notion of *sociomathematical norms*. In their documentation of a teacher intervention study in a second grade mathematics classroom, Yackel and Cobb identify the norms that regulate classroom mathematical practices. By focusing on language *in* mathematical learning, the authors argue that these norms emerge over time and are mathematical in nature; they should not be mistaken for the more generic social norms of a classroom. From the perspective of symbolic interactionism, they argue that teachers and students reflexively create these norms of mathematical work that include a shared articulation of what makes for a different or valid mathematical solution, what constitutes an efficient or sophisticated mathematical solution, how representations are used in reasoning, and so on.

In their work, they examine the language *of* mathematics by providing examples of students proffering explanations that are adjudicated discursively by their teacher and peers as a means of developing consensus on what makes an explanation valid. In one example of a socially constructed explanation, a student presents a solution to a problem using ten unit strips at the overhead projector while explaining her reasoning. In response, one of her peers, who can see how she is counting from her use of the strips, points out that the model she presents does not correspond to her own verbal reasoning. The peer goes on to argue that her claims have to correspond to what she presents through experientially real objects, in this case the ten unit strips. The authors conclude that in this way sociomathematical norms become established collectively. Moreover, when students value these norms in their own work and in evaluating the work of their peers, they become increasingly intellectually autonomous in the discourse and practices of the discipline.

Akin to Yackel and Cobb (1996), Forman, McCormick, and Donato (1998) focus on language and discourse *in* and *of* the mathematics classroom. Drawing on sociocultural theory, the authors explore how a teacher orchestrates a whole-class conversation on generalizing patterns as it relates to negotiating new norms and means of communication among the students (language *of* the mathematics classroom). At the same time, the authors illustrate how student explanations that reflect the mathematical register were more readily recognized, endorsed, and legitimated by the teacher as part of that discussion—an analysis focusing on language *in* the mathematics classroom. In their illustration, one student provides an explanation that relies on the specific patterns he built with tiles. The teacher interrupts this student and asks questions until she elicits the generalizable solution. The second student offers a form of explanation that closely mirrors the explanation modeled by the teacher when responding to the first student. The authors argue that the second student's explanation goes uninterrupted, which serves to implicitly sanction it as correct. The third and final student presents an alternate, mathematically valid solution to the problem, but the teacher does not recognize it as such because the form is altered. The teacher responds by molding the explanation to better fit with what she had intended. In this way the authors illustrate how a move from classroom discourse practices under the transmission model of communication to a model that genuinely engages students in argumentation, explanation, and other such mathematical practices, as reforms suggest, is a formidable barrier worthy of further consideration.

Forman and Ansell (2002) also examine the language *in* mathematics classrooms by focusing on the use of inscriptions, while also examining the language *of* mathematics classrooms in describing student re-voicing and teacher animation. Inscriptions—a double helix representation of DNA, a logic map—provide a public, context-specific record of thinking that is integral to discipline-based argumentation, which is seen as a central work of mathematicians and scientists. However, the authors identify a potential trap in the use of inscriptions: Students may confuse the material object and the mathematical object it presents. To address this dilemma, teachers need to develop opportunities for students to transform their understanding of the material to the mathematical. In the classroom episodes presented in this work, students connect data to inscriptions as they develop claims and arguments during a whole-class discussion. As the episode unfolds, what becomes clear to the researchers is that whereas inscriptions provide a foundation and structure for argumentation, what is also important is the ways students participate in the discussion.

In focusing on the language practices of the mathematics classroom, Forman and her co-authors pay particular attention to how talk is organized and regulated by the teacher and students in the whole-class setting. The authors draw particular attention to re-voicing (repeating, rephrasing, summarizing, elaborating, or translating) as a way for students to align with one another and the academic content of the discipline. As the data depict, students re-voice one another's claims as they develop their own arguments, while the teacher responds by animating particular explanations. The teacher's selective animation of student contributions serves to endorse and legitimize certain mathematical ideas (and indirectly, students) over others. The authors conclude that an analysis of how discipline-based tools such as inscriptions are used in classrooms, in conjunction with a study of how talk is organized and regulated, allows for a better understanding of how students come to legitimately and competently participate in the discipline.

Moschkovich (2007) similarly acknowledges the social dimensions of mathematical learning as well as the significant role of discourse in learning. She demonstrates this orientation through an analysis of a third grade bilingual mathematics classroom to illustrate two features of mathematical discourse: situated meaning of words (utterances) and focus of attention. She utilizes this analysis as a means to discuss how discourse has been understood in the field of mathematics education and presents a view of mathematical discourse that is more in line with understandings in literacy studies. Taking a sociocultural and situated perspective, the author suggests that "learning mathematics is a discursive activity that involves participating in a community of practice . . . using multiple material, linguistic, and social resources" (p. 5).

Drawing on Sfard and Cole (2003) and Gee's (1996) notion of multiple discourses, Moschkovich distinguishes discourse as more than referring to language and includes symbolic expressions, objects, and communities, and she identifies four types of mathematical discourse: everyday, professional, academic, and school. Sfard and Cole previously distinguished mathematical discourse as everyday, classroom, and academic, including professional academic discourse. Professional discourse practices refer to the practices of scientists and academic and applied mathematicians. However, for Moschkovich, school discourse refers to the practices of teachers and students, while academic discourse practices refer to what learners are expected to become fluent in. And finally, everyday mathematical discourse refers to practices adults and children engage in that are neither school nor professional.

Moschkovich contends that these categories can be misleading and may obscure the ways in which both professional and school mathematics also could be one's everyday practice. Nonetheless, these distinctions serve the purpose of exploring these discourse practices relationally. Whereas there is an emphasis on what does or does not constitute these varying discourse practices, what is important is to explore the way in which students participate in activities that are deemed "everyday" and yet that are actually "closer to the practice of scientists and mathematicians than to conventional school practices" (2007, p. 17). Thus, Moschkovich suggests that labeling the complicated ways students engage multiple resources and experiences in learning mathematics, as either academic or everyday, is no minor analytic endeavor in research.

We will explore this notion of everyday and mathematical discourse in more detail later, but first we address how this focus on mathematical discourse, as both Moschkovich and Sfard point out, is more complex than is presented both in the research literature and in mathematics education reform frameworks. The move from conceptions of mathematical discourse as a singular, monolithic discourse to understandings of multiple forms of mathematical discourse helps to illuminate a key limitation in mathematics reforms (for example the NCTM standards) relative to the socialization of students into and through mathematical discourse. There is a tendency in the literature to equate classroom discourse with academic discourse. Instead, as our own work has shown, classrooms are hybrid contexts with hybrid discourses (Gutiérrez, Baquedano-Lopez, & Tejeda, 1999; Gutiérrez, Rymes, & Larson, 1995). This is a crucial point, as the goal of apprenticing students into mathematical discourse may be untenable, difficult, or perhaps misguided if we do not acknowledge the hybrid nature of classroom discourse—that is, a hybrid of academic and everyday discourses (Sfard, 2000).

If reforms ask students to genuinely engage in mathematics as a discourse activity, students would need to develop the habits of mind—that is, the meta-rules that go beyond linguistic, grammatical canons of behavior akin to Bourdieu's (1999) notion of *habitus*. For Sfard (2000), one of the central constraints to students becoming *mathematicians* in classrooms is that there is a limit to what can happen in classrooms: First, students are not mathematicians, and further, the structure of schooling prohibits the construction of the *habitus* of mathematicians. Sfard illustrates this point through the use of historic and modern-day examples of students grappling with the concept of negative numbers. Because the students in her examples can find no real-world form equivalent that defines the value of a negative number, students come to accept the concept of negative numbers as a mathematical concept simply invented by someone—a concept to inherit. Her argument suggests that the limits of discourse in mathematics reform have everything to do with the differing meta-rules involved in such discourse. Meta-rules have a regulatory impact on mathematical learning, because they serve as tools that facilitate participation in the complex language games of a given community. These rules refer to the regularities that go well beyond, as well as include, other constructs commonly used in mathematics education literature: routines, patterns of interaction, obligations, participation structures, discursive practices, and so on. Meta-rules are "implicitly present in human interactions, [and so] meta-discursive rules are an unlikely object for a rational justification" (Sfard, 2000, p. 169) and their learning is usually a matter of practice. Meta-rules are also normative in nature, value-laden, and account for the preferred ways of behaving.

The work discussed thus far shows what is gained by studies that examine and characterize the language *in* and the language *of* mathematics class-rooms, as well as those studies that take up both dimensions to capture the complexity of developing mathematical discourse.

In the second part of this chapter we present specific works that we believe lend to a central understanding of how language, discourse, and embodied practices mediate mathematical learning and have organized these pieces around three main ideas: (1) viewing mathematical discourse as a cultural activity (i.e., talking mathematics is learning mathematics); (2) understanding the relation between everyday and scientific mathematical discourse practices; and (3) considering mathematics as a multimodal activity.

MATHEMATICS DISCOURSE AS A CULTURAL ACTIVITY: TALKING MATHEMATICS IS LEARNING MATHEMATICS

In literature we reviewed, we noted a wide range of meanings and uses of the terms language, communication, and discourse. At times, these meanings were implicit in the analyses, with the majority of the works tending to isolate language and privilege verbal expression. We believe this field of study would advance if more clarity were provided not only in defining its terms more explicitly, but also in describing their relation to one another. For example, although in the works we reviewed there was consistent attention to the social context of mathematics education settings, the activity-theoretic lens we employ entails much more than that, as we have described previously. For example, "context" in some cases was presented as that which surrounds, which is in contrast to views of "context" as being a constitutive part of the mathematical practices at work.

Within the general domain of mathematical discourse, we distinguish approaches that suggest that appropriating and participating in mathematical discourse is engaging in mathematical thinking. Discourse here, as Gee (1996, 1999) and others have argued, is a cultural activity. While we found much of the work we reviewed instructive, we found that several lines of work were particularly rich in their conceptualizations of the mediational role of discourse in mathematical learning. As an example, Sfard's (2007) communicational approach to mathematical learning provides a particularly robust notion of mathematical discourse, and we distinguish it from other work in mathematical education in several important ways.

We concur with Sfard (2001a) that the term "discourse" versus "language" is preferable in discussions of mathematical learning; to reiterate, in the Vygotskian sense, language is a tool, whereas discourse is a broader activity in which the tool is used. She addresses the longstanding question of the relation between thought and language in her work and distinguishes her work around this relationship. For Sfard, researchers with analytical perspectives on mathematics talk tend to hold one of two views: (1) thought and language are related but ultimately separate with language serving an auxiliary function in expressing thought, or (2) thought and language are inseparable. Drawing on a sociocultural view (Vygotsky, 1987), Sfard maintains that "attending to words and thought separately is like trying to find out the properties of water by looking at those of hydrogen and oxygen" (2001a, pp. 21–22).

In particular, Sfard makes the distinction between viewing discourse as one of many factors that shape learning and discourse as the object of the activity, that is, what is learned (personal communication, 2007). By using discourse both as the unit of analysis and as an index of mathematical learning, mathematical discourse serves as both vehicle and destination. In contrast, we find that many of the studies we reviewed use discourse data as the means to understand other phenomena (e.g., students' cognition, norm setting, or communities of practice)—that is, using discourse as a window through which to observe learning activity.

Specifically, Sfard (2001b) argues that the knowing of mathematics is synonymous with the ability to participate in mathematical discourse. From this perspective, conceptualizing mathematics learning as the development of a discourse and investigating how children learn mathematics means getting to know the ways in which children modify their discursive actions in three respects:

- mathematics vocabulary
- visual means with which communication is mediated
- meta-discursive rules that navigate the flow of communication and tacitly tell the participants what kind of discursive moves would count as suitable for this particular discourse, and which would be deemed inappropriate. (Sfard, 2001b, p. 26)

Appropriating mathematical discourse is fundamental in the learning process, as "becoming a participant in mathematical discourse is tantamount to learning to *think* in a mathematical way" (Sfard, 2001b, p. 26). Sfard's (2007) *commognitive approach* is grounded in the assumption that thinking is a form of communication and that learning mathematics is learning to modify and extend one's discourse.

Across her work, Sfard (2000, 2001b) discusses the chicken-and-egg dilemma around mathematical discourse—that is, the teacher needs to use mathematical terms in order to initially explain mathematical terms. Sfard addresses this complexity and claims that this circularity is what actually helps develop discourse. However, her communicational approach to cognition does not deny that people acquire mental models and schemata, but at the same time, she prefers to base her claims on data that document observable actions and interactions.

For Sfard (2001b), "learning is nothing else than a special kind of social interaction aimed at the modification of other social interactions" (p. 25). From this perspective, teachers can help modify students' everyday discourse into a more mathematical discourse, one that follows meta-discursive rules. For example, in mathematics, shapes are analytically classified by their properties, not just by how they appear to us holistically. Thus, a stretched out triangle is still a triangle even if it looks distorted. If it has three line segments joined at vertices, it is a triangle; and because we count those segments and vertices, we engage in a linguistic act.

This communicational approach to mathematics comes into contention with the "learning with understanding" movement that asserts that students must first encounter a mathematical idea, use it, and then formalize it later into mathematical conventions. While this seems like common sense, Sfard (2001b) argues that under her communicational approach to mathematics, the introduction to formal mathematics signifiers (e.g., negative numbers) is the beginning of learning not the end point. This is an inevitable part of the process, as "in order to initiate children to a discourse of new objects, one already has to use this discourse" (p. 28). She offers a two-stage process of discourse development: Stage 1 is "template-driven," where a mathematics learner inserts a mathematical idea into his or her existing understanding of that signifier; Stage 2 develops this initial sense-making into a more "objectified use of symbols":

> But herein lies a fundamental paradox: If mathematical objects, such as negative numbers, are discursive constructions, we have to talk about them in order to bring them into being. On the other hand, how can we talk about something that does not yet exist for us? ... This circularity, although an infallible source of difficulty and a serious trap for the newcomers to a discourse, is in fact the driving force behind this discourse's incessant growth. (p. 29)

She illustrates this tension by showing how a formal mathematics proof justifies why a negative number times a negative number equals a positive. However, because most students do not reason easily with formal logical proofs, this discourse is somewhat unconvincing for students as a basis to develop their mathematical discourse in the area of integers. If students try to form an "everyday" example of this concept, they come up short because their examples usually treat negative as labels for values (e.g., 5 degrees below zero), not as numbers themselves—a move that is vital to mathematical understanding. For Sfard, therefore, this illustrates a quandary in which teachers must assume the role of initiating mathematical ideas while students need opportunities to use these ideas even before they fully understand them.

Everyday and Scientific Mathematical Discourse Practices

Of significance to the discussion elaborated in this chapter is the distinction made in the field (e.g., Sfard, 2001b; Sfard & Cole, 2003) between everyday (or colloquial, or primary) and literate (or scientific, or secondary) discourses and their relationship to mathematical literacies. For example, Sfard (2001b) argues that everyday discourse does not naturally evolve into a mathematical discourse; nor are there precise mathematical analogs in the everyday world. To use her terms, the literate discourse develops from the colloquial. Specifically, Sfard argues for the co-emergence of discourse where literate and colloquial are seen as "two legs which make moving forward possible due to the fact that at any given time one of them is ahead of the other" (p. 29).

For Sfard and Cole (2003), literacy is the ability to use secondary discourses (Gee, 1991). Moreover, literate mathematical discourses are mediated by symbolic artifacts designed to communicate specific conceptual understandings of quantities. Thus, symbolic mediation is mathematical discourse's defining characteristic. Unlike spontaneously emerging everyday discourses—that is, discourses that are a part of people's routine lives—secondary discourses require deliberate teaching (Sfard & Cole, 2003, p. 3). From a communicational psychological framework, mathematics is *a* special type of discourse that involves the use of mathematical objects such as quantities and shapes. Sfard, with Cole, further distinguishes mathematical discourses to include everyday (colloquial) mathematical discourses and literate mathematical discourse—a secondary discourse that is the object of schooling.

For Sfard and Cole (2003), the meta-discursive rules in mathematics are not necessarily logical, so they cannot be intuited. To return to the example of negative numbers, in mathematics we can accept that an invented property like "negative times negative is positive" because it does not violate our number system thus far, and in fact, it expands what we can do with numbers. However, it is very unlikely that students will "discover" this meta-rule without it being taught explicitly. In other words, everyday discourse is an important starting point; however, to develop mathematical discourse requires a fundamental change in the discourse practices themselves. In short, a new discourse emerges. As Sfard notes, "If learning mathematics means an initiation to a special type of discourse, staying within the confines of everyday discourse would contradict our aim!" (2001b, p. 35).

Herein lies a central tension in mathematics education: how to reconcile the relation between the two.

Our read of the literature on everyday and scientific categories, though referenced earlier, falls within a tradition in which everyday and scientific concepts are viewed relationally. We elaborate this point as it is contrasted with views that delineate the two. The notion of everyday and scientific concepts is integral to our understanding of colloquial and secondary mathematical discourses. As we read relevant studies, we discussed the importance of unpacking this contrast in order to understand better what constitutes the difference between the two categories of everyday and secondary mathematical discourses. In particular, we were interested in whether these discourses are better represented by a continuum from protoform to scientific instead of as an implied or explicit hierarchy—that is, as representing a mutually constituted process in which scientific concepts (and discourses) grow down into the everyday (Tuomi-Grohn & Engeström, 2003).

As Scribner (1983/1997) elaborated in her work, everyday and scientific discourses are mediated by different tools, e.g., visual and symbolic. Others, such as Cole (1996), Nunes, Schlieman, and Carraher (1993), and the previously discussed work of Sfard and Cole (2003), employ and elaborate Scribner's work in the context of mathematical learning. As Sfard and Cole describe, "The quantitative discourse (and thus thinking) of the experienced workers was visually mediated by readily available familiar concrete objects, rather than by symbolic artifacts" (p. 4). We also learn from this body of work that literate mathematical discourses developed in schools are not endowed with built-in advantages over the everyday mathematical discourses developed spontaneously through repetitive practices (p. 6). But these discourses are not mutually exclusive; everyday and scientific concepts are interdependent and mutually influential (Vygotsky, 1987).

Mathematics education researchers have defined math-related language learning in a variety of ways, which makes it more difficult to achieve coherence within the field. Some define mathematics as a register (Ferrari, 2004; Pimm, 1987, 1991), which requires a different understanding of word use and forms of argument, explanation, proof, and so on. Different authors use the distinctions in discourse for different purposes: Some do it in a way that maintains a hierarchical divide between the discourses, while others use it simply to contrast them (Moschkovich, 2002, 2007). For the former group, the distinction serves to keep body separate from mind—privileging what we do with our minds over what we do with our bodies, although this position belies the very way we engage our mind and body in learning.

In contrast, Scribner's (1983/1997) investigation of dairy warehouse workers provides an instantiation of how mind and body serve learning. Scribner observes workers who load dairy products of differing sizes onto trucks in accordance with specific customer orders in order to understand their approach to problem solving. To contrast, Scribner asks a group of ninth grade students and clerks (whom she deems "novices") to engage in the same mathematical task. Both groups are observed closely in problem solving to reveal several important differences, among which is that the students tend to be single algorithm problem solvers, drawing upon a range of school-based mathematical practices, while the workers use arithmetic short-cuts in combination with the visual display. Contrasting the novices' use of symbolic tools associated with the abstractions of mathematics and the workers' visual/symbolic tools associated with the practices of the body offers a vision for mathematical learning that must at once embrace the body and mind in its accomplishment.

We were drawn to understandings of everyday and scientific discourses that blurred such categorical distinctions, specified their differences, and accounted for multimodal learning activity, including the embodiment of learning. Of course, subjugating the body to the mind is not about the truth of mathematics per se; it is as much a cultural remnant of Enlightenment as the cultural privileging of certain ways over others. Nevertheless, there is a good and useful reason to contrast these discourses, because it focuses educators on the specific competencies that students are expected to acquire in this process. By making discourse an object of attention in teaching, we make the potential of students' mastery of the domain that much better. By attending to learning as an embodied activity, we hope to help lead researchers and educators to a recognition and appreciation of the mind in the work of *applied mathematicians*: warehouse workers, street vendors, and so on, and students as they come to know mathematics.

Common to almost all of the articles in this section is a focus on the function and purpose of language as one way to draw the distinction among categories of discourse. For example, Ferrari (2004) seems to prefer a binary (discrete) distinction in the discourses rather than a continuum. Like Ferrari, McNair (1998) and O'Connor (1998) draw the distinction between the discourses as a way to understand student "failure". However, McNair and O'Connor extend their argument to include student innovation, creativity, and imagination, something not addressed in Ferrari's framework. In contrast, Moschkovich (2002) directly addresses this issue and uses these labels not to juxtapose or contrast but rather to explore their relationship.

We found work that did not dichotomize the discourses or view everyday discourses as simply serving as a bridge to secondary or literate discourses most generative. We believe that work that examines secondary discourse as a new discourse that develops from the everyday, where the everyday is also not a monolithic discourse, represents an important line of inquiry. Following Cole (personal communication, 2007), we believe these discourses are different but in specified ways, and that the acquisition of a secondary discourse is providing students an extra tool to put into their toolkits to use in the world, as we have argued, expanding students' repertoires of practice (Gutiérrez & Rogoff, 2003). Here Cole's notion of heterogeneous discourses living side by side—that is, used differently in contexts and circumstances where they are useful—is key. In other words, one discourse does not replace another (Cole & Subbotsky, 1993). Scribner and Cole (1981) similarly argued this point in their study of logical syllogisms among literate and non-literate members. They found that one mode of thinking did not replace another; rather, the modes existed side-by-side, being evoked by different circumstances for different people (Cole & Subbotsky, 1993, p. 106).

One of the dilemmas of merging the everyday and the scientific discourse is played out in the work of Godfrey and O'Connor (1995). Their work documents a teaching event in which sixth grade students are asked to create a nonstandard unit of measurement to characterize one another's height. What results is a conflict that highlights the ways in which students must become fluent in the conventions and historical grounding of mathematical terms, while also being encouraged to actively interpret and innovate on such terminology in contextually defined circumstances. In this way the coexistence of iconic terms and symbols with potentially arbitrary or contextually negotiated ones becomes a site of conflict generative for exploring complexities in student learning and teaching.

Godfrey and O'Connor (1995) suggest that engaging students in activities where they generate nonstandard measures, symbols, and notations ultimately prepares them to deal with mathematical discourse and practices as communicative acts. From our sociocultural perspective, this relationship (sometimes a tension) between conventions and improvised tools—in this case, the units of measurement—is a productive and essential one, illustrating how practices are both inherited and also potentially transformed by participants in the larger activity of mathematics.

As others have noted, the modal mathematics classroom does not privilege the kinds of communication and authorship that has characterized the history of the discipline of mathematics. Similar to Schoenfeld's (1992) notion of the "mathematical point of view," Godfrey and O'Connor (1995) note, "Most school settings do not support students coming to view themselves and their peers as communicators in the field. Thus, it is difficult to get students to actively use language and symbols in ways that will provide them with a first-person point of view on mathematical communication" (p. 340).

As part of the larger body of work investigating the contrast of in-school and out-of-school mathematics (a binary we find problematic), the work of Carraher, Carraher, and Schliemann (1985) provided some of the groundwork for rethinking the role of context in learning mathematics. The results of their empirical study on the mathematical thinking of young street vendors in Brazil revealed that children performed remarkably well at reasoning mathematically when located within a sense-making and relevant activity such as commercial trading of their products. Addition, subtraction, multiplication, and division were performed fluidly and virtually error-free without the use of pencil and paper or traditional algorithms. In contrast, when participants were given the same mathematical problems to solve using school algorithms, the children could not demonstrate the same degree of proficiency.

The results indicate that children were overwhelmingly more capable at solving problems in the informal setting, often getting waylaid by algorithms or routines associated with school mathematics during the formal test. This and similar studies have often been used as a call for "hands-on" learning or wrapping every mathematical problem in a veneer of "real-world" application, but these authors are clear in their analysis that the material artifacts (the coconuts that the children sold) did not facilitate problem solving in the informal setting in this case. However, as the authors argue, "The presence of concrete instances can be understood as a facilitating factor if the instance somehow allows the problem solver to abstract from the concrete example to a more general situation" (p. 25). While this view holds that abstract learning (as in schools) prepares one to transcend specific contexts (e.g., street vending), this work points to the richness and fluidity of mathematical thinking while (and precisely because of being) immersed in the everyday. Although this study did not have language or discourse practices as its research targets, we draw on this work to trace some of the seminal works of everyday and scientific discourse in mathematics defined more broadly.

When contrasting everyday and scientific discourses, we often locate these in either out-of-school or in-school settings where either can be developed and used (home/school). However, studies of how mathematics is used in professional practice (such as nursing) provide yet another site for studying the role and opportunity to use scientific discourse while "on the job." Pozzi, Noss, and Hoyles (1998) posit that nursing is largely mathematized, although this mathematization is invisible in its tools and daily procedures; embedded mathematical models, however, come under question when there is a breakdown in the routine. Rather than debating whether or not the work of nursing involves the doing of mathematics, the authors instead assert that we should be asking what numbers and graphs represent to nurses and how they are used. The researchers assert that the patient's chart, for instance, embodies the patient's conditions and information needed by nurses to carry out their duties. However, because the charts were made by others, they serve as tools for providing patient care without having nurses engage in any substantive mathematical thinking.

In two of the activities highlighted in their study, one (dosage administration) remains highly routinized and efficient, obviating the need for mathematical reasoning. The other practice (fluid balance monitoring), however, is interrupted when a newcomer questions a charting practice by pointing out a flaw. Through joint productive activity with a colleague, the two nurses effectively restructure the ways patient readings are charted. The authors point out the importance of linking mathematical tools and language in use. "If representations [such as the patient chart] are crucial to non-routine practice, they require a language and a set of intellectual tools which can reliably serve to express the representation not only to oneself, but to other people" (Pozzi, Noss, & Hoyles, 1998, p. 118).

Whereas others have investigated the use of mathematics in contexts outside of schooling (street selling of candies, nursing), Williams and Wake (2004) considered the discourse-divide and discourse-bridging that occurred between a research mathematician attempting to make sense of the mathematical-reliant practices and tools (spreadsheets) of a professional engineer at a gas plant. Part of the struggle is that the mathematician has entered into a new activity system; while the mathematical tools and ideas are familiar to both, the object and localized knowledge is not yet shared. Progress in shared understanding was made through conversation, wherein both the mathematician and the engineer find common ground in using common cultural models such as number lines. In addition, the authors note that the gestures the engineer uses in explaining with the timeline are essential supports for his verbal discourse. Thus, the authors push for a more complex accounting of the discourses involved in these interactions. The social languages they define include workplace language, workplace discourse genre, formal academic genres, everyday language, and gestures. The social accomplishment was a chain to connect the discourse genres of workplace and research mathematics.

Utilizing a cultural historical perspective, Schmittau (2004) identifies that while the active character of students' development of mathematical concepts is shared by both constructivists and cultural historical approaches, such as the work of Davydov (1972/1990), there are some key differences relative to views of everyday and scientific concepts: "Constructivists, however, begin the instructional process from the children's pre-existent concepts while Vygotskians reorient it toward acquisition of what Vygotsky defined as 'scientific' rather than 'spontaneous, everyday' concepts" (Schmittau, 2004, p. 19).

Schmittau highlights Davydov's (1972/1990) Vygotskian-inspired approach to mathematical development. In contrast to approaches privileged in the U.S., particularly schools with large numbers of students from non-dominant communities, Davydov and his colleagues created a mathematics program that focuses on the development of students' theoretical thinking.

Specifically, Davydov's approach engages students in a series of strategically sequenced problems that demand progressively more powerful insights and methods for their solution (Schmittau, 2004, p. 20). In general, Davydov believed in the importance of grounding the curriculum in conceptual and historical analysis. He wrote:

> The child, of course, cannot independently 'acquire' what people have already attained, but he should repeat the discoveries of people in previous generations, in a particular form. With this sort of instruction the general nature of a concept should be revealed to a child—by his own activity—before the particular manifestations. (1972/1990, p. 320)

Schmittau points out that Davydov's focus on conceptual and historical analysis is conducted such that the most general is revealed first—an approach she contrasts with the constructivist approach in which students inductively get to the general from a series of concrete examples. The assumption here is that the requisite mathematical concepts have already been culturally constructed; requiring students to construct their own would be a violation of this principle (2004, p. 21). For Schmittau (1993, 2004), underlying the dilemma of the procedural-conceptual divide are issues about everyday and scientific concepts, which are best resolved by a cultural historical approach that resists the binary.

Drawing on a cultural historical perspective, Adler's (1998) work takes up similar challenges around everyday and academic forms of mathematics. Motivated by issues of equity and social justice, Adler addresses the tensions that exist in mathematics education today: tensions around equity, diversity, as well as tensions among "spontaneous, intuitive and diverse mathematizations (everyday mathematics), the mathematics of mathematicians (the discipline of mathematics) and the canonized curriculum (school mathematics)" (p. 24). Drawing on Walkerdine's (1988) analysis, which challenges the idea that students' everyday contexts can be brought into the classroom to make learning more meaningful, Adler (1988) argues that everyday notions can be used to help make connections to mathematics, but that "mathematical meanings ... have to be pried out of their everyday discursive practice and situated in a school mathematical discursive practice" (p. 28). As Walkerdine has argued, "Non-mathematics practices become school mathematics practices, by a series of transformations, which retain links between the two practices" (1988, p. 128). Utilizing a Vygotskian (1978) perspective, Adler also sees mathematical meaning as being "linked with and emergent from other concepts ... bound in with meanings of related concepts and use. Shifting into the everyday might well not be sufficient to attach the appropriate new conceptual meaning" (p. 28).

Of significance to our review, Adler (1998) brings together the relation of classroom and mathematical discourses by suggesting that students must have access to the language of mathematics, a form of academic discourse, and to scientific concepts as well as to the classrooms' sociocultural processes, including the discourses of teaching and learning. At the same time, she makes visible a central dilemma for educators: one occurring around mediation and the other around transparency. From her perspective, the teacher is caught between the dilemma of validating pupil meanings and of developing mathematical communicative competence. Here Adler argues for making language/discourse a visible resource for learning in ways that help students develop *mathematical communicative competence* (p. 30), while validating their intuitions and informal ways of expressing their mathematical ideas. Similar to Cobb, Wood, and Yackel (1993) who distinguish between "talking about mathematics" and "talking about talking about mathematics," Adler refers to this discourse practice as "talking within and talking about the practice" (p. 30). In light of the changing demographic in U.S. schools, we believe Adler takes up a critical question in mathematics education: the need for equity-oriented mathematical programs that are also robust mathematically.

Challenging Discourse Dichotomies and Classical Views of Transfer

In discussing everyday and scientific discourses, we resist approaches that cast their relationship as separate and hierarchical. Traditional views of these discourses would argue for learners to reach beyond the everyday (primary) to a more abstract, cerebral rendering of mathematical understanding as expressed solely through its scientific (secondary) discourse. In our estimation, casting these discourses hierarchically so as to value one over the other (in this case, scientific over everyday) misjudges their relationship. We assert instead that abstract, scientific discourses come alive through particular, everyday discourses, and in this way, the relationship between discourses is interpenetrating. Therefore exploring students' understandings between everyday and scientific mathematical concepts begins with an exploration of the relationship between scientific and everyday discourses.

Rendering the two discourses as separate entities is perhaps most readily done through the commonplace notion of "transfer" which, according to Beach (2003), "invokes the metaphor of transporting something from one place to another . . . the learning from one task later applied to learning a new task" (p. 39). Some take up this idea of leveraged learning across contexts as transfer, while others see it as a "relational and interdependent process." (See Beach, 1999, 2003, for a review of notions of transfer, including views from a cultural historical perspective.) How one conceptualizes transfer is related to how one conceives of the relationship between everyday and academic discourses. We rely, in particular, on Tuomi-Grohn and

Engeström's (2003) summary of how conceptualizations of transfer differ across disciplinary and theoretical perspectives. By employing an activity theoretic perspective, we resist classical notions of transfer to view the domains relationally as Tuomi-Grohn and Engeström assert: "Scientific concepts grow down into everyday practice, into the domain of personal experience, acquiring meaning and significance, and they facilitate the mastery of the more advanced aspects of the everyday concepts" (p. 205).

In line with this and similar works (e.g., Engeström, 1999), our own work (Gutiérrez, 2008) explores the kind of expansive learning that increases what Engeström calls "vertical expertise": that is, the deepening of knowledge and practices in the disciplinary area; as well as "horizontal expertise" that requires sideway trajectories and forms of mediation between the scientific and experienced concepts (Engeström, 1999, pp. 23–24; Engeström, 2001; Tuomi-Grohn & Engeström, 2003, p. 35). We (Gutiérrez, 2008) focus on what takes hold as people move across activity settings (e.g., home, school, the corner, the virtual world) and the expertise that is developed through reaching across and crossing over boundaries, including other disciplines. Expanding the notion of learning to include horizontal expertise better accounts for students' repertoires of practice as developed across vertical and horizontal pathways of learning, including what Beach (2003) calls "consequential transitions."

By exploring the relational and mutually informing relationship between discourses, viewing the everyday and scientific relationally, and by challenging classical conceptualizations of transfer, future studies could examine how to leverage students' expansive toolkits in mathematical learning across contexts and practices. Such studies could address important questions about learning: What happens when students move from one community of practice to another, solving problems and engaging in tasks that seem new and unfamiliar? What forms of horizontal and vertical expertise will they need to develop mathematical acumen and habits of mind? These and other such questions are a call for more empirical work, especially as students' experiences are not confined to home and school alone but include movement across borders and within new social, cultural, and institutional practices.

MATHEMATICS AS MULTIMODAL ACTIVITY

Mathematics as Embodied

If we are interested in capturing the full nature of mathematical learning, new studies must consider how thinking and learning are realized in a range of embodied and multi-semiotic practices and tools. Embodiment in mathematical learning manifests in several ways, for example, as mathematical practices and as tools (which are both material and ideational). Embodied mathematical practices include hypothesizing, representing, proving, arguing, justifying, generalizing, explaining, and so on, while embodied mathematical tools include conceptual metaphors, gestures, tables, charts, symbols, inscriptions, number lines, and other representations of mathematical concepts.

In the field of embodied cognition, Nuñez et al. (1999) document the ways learning and cognition are situated and context-dependent. The researchers focus on conceptual metaphor to illustrate the embodied and socially situated nature of mathematical learning. Within this approach, a situated examination of mathematics must include an accounting of the linguistic, social, and interactional factors in the learning of mathematics (p. 45). We look to the works of Lakoff, Nuñez and colleagues (Lakoff & Nuñez, 1997, 2000; Nuñez et al., 1999; Nuñez, 2006), who use techniques such as cognitive semantics and gesture studies to document how mathematical learning is ultimately embodied in nature. Their work is instructive, as we learn that "via everyday human embodied mechanisms such as conceptual metaphor and conceptual blending, the inferential patterns drawn from direct bodily experience in the real world get extended in very specific and precise ways to give rise to a new emergent inferential organization in purely imaginary domains" (Nuñez, 2006, p. 162).

Using gesture studies to examine the instructional discourse practices of mathematics professors, Nuñez et al. (1999) were able to document that the instructors were not only using metaphorical linguistic expressions, but they also were 'thinking dynamically.' Thus, even the most abstract and precise mathematical ideas are ultimately embodied. This work helps us understand how the notion of embodiment is taken up across cognitive science, including cognitive linguists (Lakoff & Johnson, 1998; Lakoff & Nuñez, 1997) and the relation between cognition, mind, and embodied experience in the world.

Of relevance to the discussion in this chapter, gestures show the dynamism involved in mathematical ideas and help us understand the limitations of the formal mathematical register.

An embodied cognitive view of mathematics education is important because it helps capture the richness of mathematical ideas. As Nuñez and colleagues (1999) argue,

> The conceptual structures which emerge in the human mind to make sense of our bodily experience provide the raw material for the construction of shared communication through language, and subsequently, the shared construction of meanings. Thus, our understandings of the world, and of mathematics, may be socially and culturally situated, but it is the commonalties in our physical embodiment and experience that provide the bedrock for this situatedness. (pp. 62–63)

Within mathematics education, Godfrey and O'Connor (1995) provide an especially rich illustration of mathematical learning as an embodied practice in their work in a sixth grade mathematics class. As we have previously mentioned, this study describes a teaching episode in which students are discussing their creation of nonstandard units to measure one another's height. Engaging students in activities where they generate nonstandard measures, symbols, and notations prepares them to deal with "traditional mathematical texts and language as real communication. These activities are thus intended to provide a way into a community that creates new measures and symbols as they are needed" (p. 328).

As the event unfolds, students engage in the authentically mathematical practice of negotiating definitions as they argue over how best to use "handspan" as a nonstandard unit of measure. One student, Kadeem, argues the handspan should be measured vertically, from heel of palm to tip of finger, while Elliot insists "handspan" refers to the horizontal distance from tip of thumb to tip of pinky. Elliot eventually turns to a dictionary as a linguistic authority codified by history to justify his claim and refute Kadeem's argument to suspend such conventions. Kadeem, however, expresses a position that is mathematically valid: terms may be contextually negotiated and defined.

In a related discussion, students object to Kadeem's use of a six-fingered hand with a vertical arrow as the symbolic representation of his handspan. The students argue that such a symbol would have little meaning outside of their classroom community. Kadeem, however, steadfastly insists that his symbol and nonstandard unit of measurement are appropriate, effectively balancing iconicity and arbitrariness. The students eventually resolve the situation by developing two terms: a "handspan" to refer to Elliot's conception and symbolized by a five-fingered hand and a "Kadeem handspan" to refer to the vertical handspan symbolized by a six-fingered hand. Embodiment, then, encompasses different kinds of phenomena.

Mathematics as Multi-Semiotic

We turn to linguistic anthropology to understand better what is meant by the multi-semiotic nature of human activity. Goodwin's (2000) notion of cognition as a "reflexively situated process" (p. 1490) is particularly useful as it highlights "the sign-making capacity of the individual... through the production of talk, and different kinds of semiotic phenomena, from sequential organization to graphic fields lodged within the material and social environment" (p. 1490). Goodwin's emphasis on cognition as a social process that is an inextricable part of the historically shaped material world is highly aligned with the cultural historical activity theoretic approaches and social and anthropological studies of scientific and workplace practice we emphasize in this chapter.

Specifically, we are interested in those semiotic domains involved in mathematical learning, particularly because a focus on semiotic mediation helps us understand that language is not an autonomous system and instead is a social accomplishment in discourse. Thus, rather than conceiving of language as the outcome of private psychological processes situated within a single individual, Hasan (2002) argues that language "is a form of public practice lodged within the organization of action within human interaction" (p. 98). Semiotic mediation is a key notion in a sociocultural approach as it emphasizes the fundamental relationship between human mental activities and discourse in sociocultural activity (Vygotsky, 1978; Wertsch, 1985). Put simply, the term *semiotic* refers to all modalities for signing, not just language (Hasan, 2002). As Hasan has noted, "Identifying sociocultural activity as the essential site for the operation of semiotic mediation emphasizes the relation between cultural activities and language, and whether different kinds of activities encourage different forms of semiotic mediation" (p. 2).

Interestingly, there are few studies that take up the semiotics of mathematics. For example, O'Halloran (2005) has elaborated the notion of mathematics as a "multi-semiotic" activity. Specifically, O'Halloran maintains that multi-semiotic constructions are "discourses formed through choices from the functional sign systems of language, mathematical symbolism, and visual display" (p. 10). From a multi-semiotic perspective, written modes, spoken discourse, physical action, gestures in the environment, including digital media and three-dimensional material reality, are multi-semiotic resources (O'Halloran, 2005).

Research from this perspective, O'Halloran argues, may have particular pedagogical affordances, insofar as teachers and students are not aware of the grammatical systems for mathematical symbolism and visual display that are found in mathematical texts and classrooms (O'Halloran, 2005). In particular, a systemic functional approach to mathematics as a multi-semiotic discourse provides a framework that explains how language, mathematical symbolism, and visual images (such as graphs and diagrams) function *intersemiotically*. Specifically, mathematical discourse is successful because:

1. The meaning potentials of language, symbolism and visual images are accessed.
2. The discourse, grammatical and display systems of each resource function integratively.
3. Meaning expansions occur when the discourse shifts from one semiotic resource to another. (p. 204)

Thus, a systemic functional approach to language provides students a model for understanding language as a tool and meaning as a choice (p. 200). Here the grammar of mathematical symbolism differs from the grammar of language and thus needs to be made explicit in educational contexts. This is particularly important as in many educational contexts today the mathematics curriculum is pre-packaged; hence the functions and grammar of mathematical symbolism are neither evident nor discussed. Finally, the grammar of visual images needs to be learned in order to understand the specific systems that link to symbolism (p. 203). (For other examples of work that take up the semiotics of mathematics, see Anderson, Saenz-Ludlow, Zellweger, & Cifarelli, 2003; O'Halloran, 2000; Rotman, 1988).

Of relevance to a multimodal understanding of mathematics, semiotic domains are sets of practices that "recruit one or more modalities (e.g., oral or written language, images, equations, symbols, sounds, gestures, graphs, artifacts, etc.) to communicate distinctive types of meanings" (Gee, 2003, p. 18). This is a particularly important point in relation to the development of domain literacies, including mathematical literacy, as people read and write differently across domains. As Gee notes, "Each requires its own rules and requirements...each is a culturally and historically separate way of reading and writing, and, in that sense, a different literacy. They are also related to social practices" (p. 15). In line with this thinking, he argues, "Knowing about a social practice always involves recognizing various distinctive ways of acting, interacting, valuing, feeling, knowing, and using various objects and technologies that constitute the social practice" (p. 16).

This understanding of learning is compatible with research studies in mathematics education that most interested us—scholarly work that viewed mathematics as historically and culturally organized and mediated through a range of artifacts and contexts. As we discussed previously in this chapter, this body of work emphasizes the importance of making mathematical discourse explicitly available, as well as making the social practices the object of reflection and meta-cognitive awareness.

In our own work (Gutiérrez, 2008), we have documented how developing a learning ecology organized around cultural historical principles of learning, mediated by a range of discourse and embodied practices, fosters the development of a semiotic toolkit for deep learning across subject matter domains with particular affordances for students from non-dominant communities. Essential to the students' learning is making visible the social practices in which they participate as well as the tools that mediate their understandings. Gee (2003) has termed such learners "producers" and discusses their paradox in ways that we believe are similar to emerging arguments in communicative and cultural historical approaches to mathematical thinking and learning:

[T]here is a paradox about producers. On one hand, producers are deeply enough embedded in their social practices that they can understand the texts associated with those practices quite well. On the other hand, producers are often so deeply embedded in their social practices that they take the meanings and values of the texts associated with those practices for granted in an unquestioning way. (Gee, 2003, p. 16)

For Gee, deep learning requires producer-like learning and knowledge and critical reflection. How will schooling environments engender such practices?

DEVELOPING A MATHEMATICAL VISION

We believe that mathematical learning involves developing what we term a "mathematical vision" (Gutiérrez, Sengupta-Irving, Dieckmann, 2006). Paraphrasing Goodwin's (1994) notion of a professional vision and Gee's (1996) idea about knowing a practice, a mathematical vision is a socially organized way of seeing, understanding, envisioning, and doing mathematics in ways that are accountable to the distinct norms of the mathematical community (Goodwin, 1994; see also Yackel & Cobb, 1996). Seeing mathematically is not a transparent, intrapsychological process but instead, according to Goodwin, "is a socially situated activity accomplished through the deployment of a range of historically constituted discursive [and embodied] practices" (1994, p. 607).

Within this more robust and ecological perspective, a mathematical vision is accomplished through discursive practices in which tools, images, diagrams, inscriptions, talk, the body, standards and disciplinary norms, and the social context of activity play a central mediating role in mathematical learning. Of significance, the social practices and activities that constitute seeing mathematically are embedded within specific communities, and they must be made visible and learned. Thus, to become mathematically literate requires explicit attention to developing an expansive linguistic and artifact-rich repertoire, including the sociocultural knowledge of what it means to do mathematics.

The work of Cobb, Boufi, McClain, and Whitenack (1997) takes up Gee's call and highlights critical reflection as being integral to deep learning by examining reflective discourse—that is, repeated shifts from what students do in action to what becomes an explicit object of discussion. In a year-long study of a first grade classroom, students' sensory-motor and conceptual activity were observed as the source of ways of knowing and meaningful mathematics activity as creating and manipulating experientially real mathematical objects. Through close analyses of student–teacher interactions, the researchers observed that opportunities for students to reflect individually and collectively occur when teachers guide and shift the discourse to make students' mathematical actions an object of study.

Burton (1998) relays a professional vision of mathematics based on interviews with seventy European mathematicians. Through an analysis of their individual "life histories," Burton offers a theoretical model for what it means to know mathematics that involves the following: (1) person and cultural-social relations, (2) aesthetics, (3) intuition and insight, (4) different approaches, and (5) connectivities. This work unearths a vision of mathematics that is at once collaborative, community-based, cultural, and even emotional in nature. Burton concludes that it is impossible to talk about mathematics practices as uniform and mathematicians as discrete from these practices. Rather, coming to know mathematics is envisioned as membership in a community and a natural outcome of joint, cultural activity that cannot be separated from the people who create it. As Livingston (2006) maintains,

> [In] studies of domains of mundane expertise such as mathematics, chemistry or checkers, chief among these techniques and essential to all of them is the serious engagement in the tasks of learning the skills of that domain. A novice is in a position to see and attend to what is most ordinary and, therein, transparent to experienced practitioners. (p. 65)

Livingston's (1999) ethnography of the "cultures of proving" illustrates the agreed-upon practice among mathematicians to display proofs publicly, not as plodding logical deductions, but through inventive representations that encode their reasoning and with which only discerning readers can delight and engage. He points to the mathematics community's preferences to have proofs *appear* transcendental as though they were available prior to the mathematical argument. They are meant to appear self-evident. He notes, "When provers arrange, re-arrange, and re-work the material details of a prospective and developing proof, they are, in fact, orienting to and composing the cultural substance of their work" (p. 880). Most mathematicians work diligently to erase the human hand from their products to give the impression that mathematical ideas and relationships are *revealed* rather than constructed. From our perspective, therefore, mathematical vision is binocular: One ocular readily acknowledges the human hand as intervening and inventing and the other more public ocular that presents these works as derived from logic alone.

LOOKING TOWARD THE FUTURE

Following Davydov's (1988, 1972/1990) concept of ascending from the abstract to the concrete, we turn our focus to the material consequences on students of conceiving of mathematics as a cultural activity and the mediating role language plays in mathematical learning. We have discussed the importance of attending to the language of the mathematics classroom (the discourse practices that constitute the everyday practices in the classroom) and the language in mathematics activity that helps constitute mathematical practices. Recognizing that students are socialized to mathematical discourse through their participation in the discourse practices of the mathematics activity is particularly useful in teaching mathematics to students from non-dominant communities, including dual language learners who are in the process of appropriating academic discourse and the discourse of mathematical thinking.

In applying a cultural historical theoretical perspective, we have also noted the importance of moving away from deficit understandings of students' knowledge and practices to more robust understandings of the relation of everyday knowledge to school-based or scientific knowledge. We have addressed this in two ways. First, we believe that a focus on understanding and "exploiting" students' history of involvement with mathematical practices across time and space—their repertoires of mathematical practices—is more productive for developing expansive teaching and learning and for studying what students can do. This approach is simultaneously historical and proleptic, insofar as it leverages students' repertoires towards future learning, principally robust mathematical understandings and practices. In this way, we bring together horizontal and vertical expertise (Engeström, 2001; Gutiérrez, 2008) and thus exploit the relations between everyday and scientific knowledge in ways we have discussed in this chapter.

Further, the discussions of the various discourses of mathematical classrooms (e.g., *in* and *of*) and of everyday and scientific mathematical practices advanced here should serve as a call for mathematics education researchers to recognize the importance of understanding everyday and scientific discourse practices, as well as everyday and scientific mathematical understandings relationally in order to expand students' semiotic toolkit and extend their repertoires of practice. This is of critical importance for English learners and students from non-dominant communities for whom these discourse practices may be less familiar. Thus, developing mathematical discourse is both an academic and an equity issue.

At the same time, while we make use of a cultural historical activity theoretic approach to learning in this chapter, we have been mindful of the importance and centrality of the discipline of mathematics itself in this endeavor; that is, mathematics resides at the heart of this learning. Our chapter takes seriously the cautionary words of Thames and Ball (2004) who remind us about the ways a focus on learning can obfuscate attending to the mathematics enterprise:

In many ways, attention to a theory of learning is the culprit...in the field at large. A theory of learning, sociocultural or otherwise, is less likely to hold onto the mathematics of teaching and learning. Research built solely on theories of learning will repeatedly turn our attention back to the psychological and sociological basis for classroom interaction, rather than the mathematical. (p. 430)

The stance we take does not view the mathematics classroom or mathematics itself as a context that surrounds, but rather the classroom and mathematical practices as mutually constitutive of the very practices and tools of mathematical learning.

Taken together, these are the issues that we believe are fundamental to conceptions of and practices of teaching and learning mathematics robustly. Equity, from this perspective, is linked to supporting students' opportunities to engage in powerful mathematics learning that is organized around expansive notions of learning and development, and additive rather than subtractive schooling practices. How would trajectories of mathematics learning shift if we leveraged students' everyday knowledge and practices? If we understood their relation to deeper forms of learning? This is the challenge for mathematics education researchers and practitioners.

NOTES

1. This review was conducted by Gutiérrez in 2005–2006.
2. Gutiérrez interviewed scholars in person or by email. In particular, we were interested in their understanding of the role of language and embodied practices in learning mathematics. This chapter reflects our views and unless otherwise specified should not be attributed to the interviewees (Na'ilah Nasir, Shelley Goldman, Jo Boaler, Daniel Schwartz, Anna Sfard, and Judit Moschkovich).
3. We use the term *transfer* recognizing that the construct comes from a particular theoretical tradition in learning in which knowledge used in one task is used to solve a problem in another task (Thorndike & Woodworth, 1901). Utilizing a sociocultural approach, our own work would focus on change in individuals' actions in a range of changing practices and social organizations (Gutiérrez, 2008; Tuomi-Grohn & Engestrom, 2003).

REFERENCES

Adler, J. (1998). A language of teaching dilemmas: Unlocking the complex multilingual secondary mathematics classroom. *For the Learning of Mathematics, 18*(1), 24–33.

Anderson, M., Saenz-Ludlow, A., Zellweger, S., & Cifarelli, V. (Eds.). (2003). *Educational perspectives on mathematics as semiosis: From thinking to interpreting to knowing*. Ottawa: Legas Publishing.

Ball, D. L. (1993). With an eye on the mathematical horizon: Dilemmas of teaching elementary school mathematics. *Elementary School Journal, 93*, 373–397.

Beach, K. D. (1999). Consequential transitions: A sociocultural expedition beyond transfer in education. *Review of Research in Education, 24*, 101–139.

Beach, K. D. (2003). Consequential transitions: A developmental view of knowledge propagation through social organizations. In T. Tuiomi-Grohn & Y. Engeström (Eds.), *Between school and work: New perspectives on transfer and boundary-crossing* (pp. 39–62). New York: Pergamon.

Bourdieu, P. (1999). Structures, *habitus*, practices. In A. Elliot (Ed.), *The Blackwell reader in contemporary social theory* (pp. 107–118). Oxford, England: Blackwell.

Brown, C. A., Stein, M. K., & Forman, E. A. (1996). Assisting teachers and students to reform the mathematics classroom. *Educational Studies in Mathematics, 31*, 63–93.

Burton, L. (1998). The practices of mathematicians: What do they tell us about coming to know mathematics? *Educational Studies in Mathematics, 37*(2), 121–143.

Burton, L. (2001, August). Children's mathematical narratives as learning stories. *Keynote address presented at the European Conference on Quality in Early Childhood Education (EECERA)*. Alkmaar, Netherlands.

Carpenter, T., Fennema, E., & Franke, M. L. (1996). Cognitively guided instruction: A knowledge base for reform in primary mathematics instruction. *Elementary School Journal, 97*, 3–20.

Carraher, T., Carraher, D., & Schliemann, A. (1985). Mathematics in the streets and in schools. *British Journal of Developmental Psychology, 3*(1), 21–29.

Cobb, P., Wood, T., & Yackel, E. (1993). Discourse, mathematical thinking, and classroom practice. In E. Forman, N. Minick, & C. A. Stone (Eds.), *Contexts for learning: Sociocultural dynamics in children's development* (pp. 91–119). New York: Oxford University Press.

Cobb, P., Boufi, A., McClain, K., & Whitenack, J. (1997). Reflective discourse and collective reflection. *Journal for the Research of Mathematical Education, 28*(3), 258–277.

Cole, M. (1996). *Cultural psychology: A once and future discipline*. Cambridge, MA: The Belknap Press of Harvard University Press.

Cole, M., & Engeström, Y. (1993). A cultural-historical approach to distributed cognition. In G. Salomon (Ed.), *Distributed cognition: Psychological and educational considerations* (pp. 1–46). New York: Cambridge University Press.

Cole, M., & Subbotsky, E. (1993). The fate of stages past: Reflections on the heterogeneity of thinking from the perspective of cultural–historical psychology. *Schweizerische Zeitschrift fur Psychologie, 52*(2), 103–113.

Cole, M., Gay, J., Glick, J. A., & Sharp, D. W. (1971). *The cultural context of learning and thinking*. New York: Basic Books.

Confrey, J. (1990). A review of the research on student conceptions in mathematics, science, and programming. In C. Cazden (Ed.), *Review of research in education* (vol. 16, pp. 3–55). Washington, DC: American Educational Research Association.

D'Ambrosio, U. (1984). *Socio-cultural bases for mathematical education*. Campinas, Brazil: UNICAMP.

Davydov, V. V. (1990). *Types of generalization in instruction: Logical and psychological problems in the structuring of school curricula*. Reston, VA: National Council of Teachers of Mathematics. (Original work published 1972)

Davydov, V. V. (1988). Problems of developmental teaching: The experience of theoretical and

experimental psychological research. Parts I – III. *Soviet Education, 30*, 8–10.

Durkin, K., & Shire, B. (Eds.) (1991). *Language in mathematical education: Research and practice*. Bristol, PA: Open University Press

Engeström, Y. (1987). *Learning by expanding: An activity-theoretic approach to developmental research*. Helsinki: Orienta-Konsultit.

Engeström, Y. (1999). *Introduction. Lernen durch Expansion*. (F. Seeger, Trans). Marburg: BdWi-Verlag, Germany.

Engeström, Y. (2001, March). *The horizontal dimension of expansive learning: Weaving a texture of cognitive trails in the terrain of health care in Helsinki*. Paper presented at New Challenges to Research on Learning, University of Helsinki, Finland.

Engeström, Y. (Ed.). (2005). *Developmental work research: Expanding activity theory in practice, 12*. Berlin, Germany: International Cultural-Historical Human Sciences, Lehmanns Media.

Ferrari, P.L. (2004). Mathematical language and advanced mathematics learning. In M. Johnsen Hoines & A. Berit Fugelstad (Eds.), *Proceedings of the 28th Conference of PME* (pp. 383–390). Bergen, Norway: Bergen University College.

Forman, E. A. (1996). Learning mathematics as participation in classroom practice: Implications of sociocultural theory for educational reform. In L. P. Steffe, P. Nesher, P. Cobb, G. A. Goldin, & B. Greer (Eds.), *Theories of mathematical learning* (pp. 115–130). Mahwah, NJ: Lawrence Erlbaum Associates.

Forman, E. A., & Ansell, E. (2002). Orchestrating the multiple voices and inscriptions of a mathematics classroom. *The Journal of the Learning Sciences, 11*(2&3), 251–274.

Franke, M., Kazemi, E., & Battey, D. (2007). Understanding teaching and classroom practice in mathematics. In F. Lester (Ed.), *Second handbook of research on mathematics teaching and learning* (pp. 225–256). Charlotte, NC: Information Age Publishing.

Forman, E., McCormick, D., & Donato, R. (1998). Learning what counts as a mathematical explanation. *Linguistics and Education, 9*(4), 313–339.

Gay, J. & Cole, M. (1967). *The new mathematics and an old culture*. New York: Holt, Rinehart, & Winston.

Gee, J. (1991). What is literacy? In C. Mitchel & K. Weiler (Eds.), *Rewriting literacy: Culture and the discourse of the other* (pp. 3–11). New York: Bergin & Garvey.

Gee, J. (1996). *Social linguistics and literacies: Ideology in discourses* (2nd ed.). London: Taylor & Francis.

Gee, J. (1999). *An introduction to discourse analysis: Theory and method*. New York: Routledge.

Gee, J. (2003). *What video games have to teach us about learning and literacy*. New York: Palgrave.

Godfrey, L., & O'Connor, M. C. (1995). The vertical handspan: Nonstandard units, expressions, and symbols in the classroom. *Journal of Mathematical Behavior, 14*(3), 327–345.

Goffman, E. (1974). *Frame analysis: An essay on the organization of experience.* New York: Harper & Row.

Goldman, S. (2005). A new angle on families: Connecting the mathematics in daily life with school mathematics. In A. Bekerman, N. Burbules, & D. Silberman-Keller (Eds.), *Learning in places: The informal education reader* (pp. 55–76). Bern: Peter Lang Publishing Group.

Goodwin, C. (1994). Professional vision. *American Anthropologist, 96*(3), 606–633.

Goodwin, C. (2000). Action and embodiment within situated human interaction. *Journal of Pragmatics, 32,* 1489–1522.

Goodwin, C. (2006). Human sociality as mutual orientation in a rich interactive environment: Multimodal utterances and pointing in aphasia. In N. J. Enfield & S. C. Levinson (Eds.), *Roots of human sociality, culture, cognition, and interaction* (pp. 97–125). Oxford: Berg Publishers.

Green, J., & Dixon, C. (1994). Talking knowledge into being: Discursive and social practices in classrooms. *Linguistics and Education, 5,* 231–239.

Gutiérrez, K. (2008). Developing a sociocritical literacy in the third space. *Reading Research Quarterly, 43*(2), 148–164.

Gutiérrez, K., & Rogoff, B. (2003). Cultural ways of learning: Individual traits or repertoires of practice. *Educational Researcher, 32*(5), 19–25.

Gutiérrez, K., Baquedano-Lopez, P., & Tejeda, C. (1999). Rethinking diversity: Hybridity and hybrid language practices in the third space. *Mind, Culture, & Activity: An International Journal, 6*(4), 286–303.

Gutiérrez, K., Rymes, B., & Larson, J. (1995). Script, counterscript, and underlife in the classroom: James Brown versus Brown v. Board of Education. *Harvard Educational Review, 65*(3), 445⁻471.

Gutiérrez, K., Sengupta-Irving, T. & Dieckmann, J. (2006). *Developing a mathematical vision: Mathematics as a discursive and embodied practice.* Chicago, IL: Spencer Foundation.

Hasan, R. (2002). Semiotic mediation and mental development in pluralistic societies: Some implications for tomorrow's schooling. In G. Wells & G. Claxton (Eds.), *Learning for life in the 21st century: Sociocultural perspectives on the future of education* (pp. 112–126). Oxford, UK: Blackwell.

Hufferd-Ackles, K., Fuson, K., & Sherin, M. G. (2004). Describing levels and components of a math-talk community. *Journal for Research in Mathematics Education, 35*(2), 81–116.

Hymes, D. (1977). Critique. *Anthropology and Education Quarterly, 8,* 91–93.

Kerslake, D. (1991). The language of fractions. In K. Durkin & B. Shire (Eds.), *Language in mathematical education: Research and practice* (pp. 85–94). Bristol, PA: Open University Press.

Kieran, C., Forman, E., & Sfard, A. (Eds.) (2002). *Learning discourse: Discursive approaches to research in mathematics education.* Dordrecht, The Netherlands: Kluwer Academic Publishers.

Kieren, T. (1999, October). *Language use in embodied action and interaction in knowing fractions.* Paper presented at the Annual Meeting of the North American

Chapter of the International Group for the Psychology of Mathematics Education. 21st Cuernavaca, Morelos, Mexico.

Lakoff, G., & Johnson, M. (1998). *Philosophy in the flesh.* New York: Basic Books.

Lakoff, G., & Nuñez, R. (1997). The metaphorical structure of mathematics: Sketching out cognitive foundations for a mind-based mathematics. In L. English (Ed.), *Mathematical reasoning: Analogies, metaphors, and images* (pp. 21–89). Mahwah, NJ: Erlbaum.

Lakoff, G., & Nuñez, R. (2000). *Where mathematics comes from: How the embodied mind brings mathematics into being.* New York: Basic Books

Lampert, M. (1998). Introduction. In M. Lampert & M. L. Blunk (Eds.), *Talking mathematics in school* (pp. 1–14). Cambridge: Cambridge University Press.

Lave, J. (1988). *Cognition in practice.* Cambridge: Cambridge University Press.

Leont'ev, A. N. (1978). *Activity, consciousness, and personality.* Englewood Cliffs, NJ: Prentice-Hall.

Lerman, S. (1996). Intersubjectivity in mathematics learning: A challenge to the radical constructivist paradigm. *Journal for Research in Mathematics Education, 27*(2), 224–250.

Lerman, S. (2002). Cultural, discursive psychology: A sociocultural approach to studying the teaching and learning of mathematics. *Educational Studies in Mathematics, 46,* 87–113.

Lin, L. (1994). Language of and in the classroom: Constructing the patterns of social life. *Linguistics and Education, 5,* 367–409.

Livingston, E. (1999). Cultures of proving. *Social Studies of Science, 29,* 867–888.

Livingston, E. (2006). The context of proving. *Social Studies of Science, 36*(1), 39–68.

McNair, R. E. (1998). Building a context for mathematical discussion. In M. Lampert & M. L. Blunk (Eds.), *Talking mathematics in school* (pp. 82–106). New York: Cambridge University Press.

Mehan, H. (1978). Structuring school structure. *Harvard Educational Review, 48*(1), 32–64.

Mehan, H. (1979). *Learning lessons.* Cambridge, MA: Harvard University Press.

Morgan, C. (1996). 'The language of mathematics': Towards a critical analysis of mathematics texts. *For the Learning of Mathematics, 16*(3), 2–10.

Morgan, C. (1998). *Writing mathematically: The discourse of investigation.* London: Falmer Press.

Moschkovich, J. N. (2002). An introduction to examining everyday and academic mathematical practices. In M. Brenner & J. Moschkovich (Eds.), *Everyday and academic mathematics in the classroom.* JRME Monograph Number 11 (pp. 1–11). Reston, VA: NCTM.

Moschkovich, J. N. (2007). Examining mathematical Discourse practices. *For the Learning of Mathematics, 27*(1), 24–30.

Nasir, N. (2000). Points ain't everything: Emergent goals and average and percent understandings in the play of basketball among African-American students. *Anthropology and Education Quarterly, 31*(3), 283–305.

Nasir, N. (2002). Identity, goals, and learning: Mathematics in cultural practice. *Mathematical Thinking and Learning,* Special issue on Diversity, Equity, and Mathematical Learning, N. Nasir & P. Cobb (Eds.), *4*(2&3), 213–248.

Nasir, N., & Hand, V. M. (2004, April). *From the court to the classroom: Managing identities as learners in basketball and classroom mathematics.* Paper presented at the American Educational Research Association, San Diego, CA.

National Council of Teachers of Mathematics. (1991). *Professional standards for teaching mathematics.* Reston, VA: National Council of Teachers of Mathematics.

National Council of Teachers of Mathematics. (2000). *Principles and standards for school mathematics.* Reston, VA: National Council of Teachers of Mathematics.

Nunes, T., Schlieman, A. D., & Carraher, D. W. (1993). *Street mathematics and school mathematics.* Cambridge: Cambridge University Press.

Nuñez, R. (2006). Do real numbers really move? Language, thought, and gesture: The embodied cognitive foundations of mathematics. In R. Hersh (Ed.), *Unconventional essays on the nature of mathematics* (pp. 161–181). New York: Springer.

Nuñez, R., Edwards, L., & Matos, J. F. (1999). Embodied cognition as grounding for situatedness and context in mathematics education. *Educational Studies in Mathematics, 39*(45), 45–65.

Ochs, E. (1988). *Culture and language development: Language acquisition and language socialization in a Samoan village.* Cambridge: Cambridge University Press.

O'Connor, M. C. (1998). Language socialization in the classroom: Discourse practices and mathematical thinking. In M. Lampert & M. L. Blunk (Eds.), *Talking mathematics in school* (pp. 17–55). Cambridge: Cambridge University Press.

O'Halloran, K. L. (2000). Classroom discourse in mathematics: A multisemiotic analysis. *Linguistics and Education, 10*(3), 359–388.

O'Halloran, K. L. (2005). *Mathematical discourse: Language, symbolism, and visual images.* New York: Continuum.

Pimm, D. (1987). *Speaking mathematically: Communication in mathematics classrooms.* London: Routledge & Kegan Paul.

Pimm, D. (1991). Communicating mathematically. In K. Durkin & B. Shire (Eds.), *Language in mathematical education: Research and practice* (pp. 17–23). Bristol, PA.: Open University Press.

Pozzi, S., Noss, R., & Hoyles, C. (1998). Tool in practice: Mathematics in use. *Educational Studies in Mathematics, 36*(2), 105–122.

Roth, W. M. (2001). Gestures: Their role in teaching and learning. *Review of Educational Research, 71*(3), 365–392.

Rotman, B. (1988). Towards a semiotics of mathematics. *Semiotica, 72*(1/2), 1–35.

Saxe, G. B. (1988). The mathematics of child street vendors. *Child Development, 59*(5), 1415–1425.

Saxe, G. B. (1991). *Culture and cognitive development: Studies in mathematical understanding.* Hillsdale, NJ: Lawrence Erlbaum Associates.

Schmittau, J. (1993). Vygotskian scientific concepts: Implications for mathematics education. *Focus on Learning Problems in Mathematics, 15*(2&3), 29–39.

Schmittau, J. (2004). Vygotskian theory and mathematics education: Resolving the conceptual-procedural dichotomy. *European Journal of Psychology of Education, 19*(1), 19–43.

Schoenfeld, A. H. (1992). Learning to think mathematically: Problem solving, metacognition, and sense making in mathematics. In D. Grouws (Ed.), *Handbook for research on mathematics teaching and learning* (pp. 334–370). New York: Macmillan.

Schwartz, D., & Black, T. (1999). Influences through imagined actions: Knowing by simulated doing. *Journal of Experimental Psychology: Learning, Memory, and Cognition, 25*(1) 116–136.

Scribner, S. (1983/1997). Mind in action: A functional approach to thinking. In M. Cole, Y. Engeström, & O. Vasquez (Eds.), *Mind, culture, and activity: Seminal papers from the Laboratory of Comparative Human Cognition* (pp. 354–368). Cambridge, U.K.: Cambridge University Press.

Scribner, S., & Cole, M. (1981). *The psychology of literacy.* Cambridge, MA: Harvard University Press.

Sfard, A. (2000). On reform movement and the limits of mathematical discourse. *Mathematical Thinking and Learning, 2*(3), 157–189.

Sfard, A. (2001a). Communicating to learn or learning to communicate? Mathematics education in quest for new answers to old questions. (Book review). *Zenralblatt fur Didiaktik Mathematik/International Reviews on Mathmatics Education, 33*(1), 17–25.

Sfard, A. (2001b). Learning mathematics as developing a discourse. In R. Speiser, C. Maher, & C. Walter (Eds.), *Proceedings of 21st Conference of PME-NA* (pp. 23–44). Columbus, OH: Clearing House for Science, Mathematics and Environmental Education.

Sfard, A. (2007). When the rules of discourse change, but nobody tells you: Making sense of mathematics learning from a commognitive standpoint. *Journal of Learning Sciences, 16*(4), 567–615.

Sfard, A., & Cole, M. (2003, April). *Literate mathematical discourse: What it is and why should we care?* Paper presented at the annual meeting of the National Council of Teachers of Mathematics, Las Vegas, NV. http://lchc.ucsd.edu/vegas.htm.

Street, B., Rogers, A., & Baker, D. (2006, April). *Adult teachers as researchers: Ethnographic approaches to numeracy and literacy as social practices.* Paper presented at the Annual Meeting of the American Educational Research Association, Chicago, IL.

Thames, M. H., & Ball, D. L. (2004). Book review of *Learning discourse: Discursive approaches to research in mathematics education. Mathematical Thinking and Learning, 6*(4), 421–433.

Thorndike, E. L., & Woodworth, R. S. (1901). The influence of improvement in one mental function upon the efficiency of other functions: (I). *Psychological Review, 8,* 247–261.

Tuomi-Grohn, T., & Engeström, Y. (2003). From transfer to boundary-crossing between school and work as a tool for developing vocational education: An introduction. In T. Tuomi-Grohn & Y. Engeström (Eds.), *Between school and work: New perspectives on transfer and boundary crossing* (pp. 1-15). New York: Pergamon.

Vygotsky, L. S. (1962). *Thought and language.* Cambridge, MA: MIT Press.

Vygotsky, L. S. (1978). *Mind in society: The development of higher psychological processes.* Cambridge, MA: Harvard University Press.

Vygotsky, L. S. (1987). *The collected works of L. S. Vygotsky (Vol. 1): Problems of general psychology* (N. Minick, Ed. & Trans.). New York: Plenum.

Vygotsky, L. S. (1997). *The collected works of L. S. Vygotsky (Vol. 4): The history of the development of higher mental functions*. (R. Rieber, Ed. & M. Hall, Trans.). New York: Kluwer Academic Publishers

Walkerdine, V. (1988). *The mastery of reason: Cognitive development and the production of rationality*. London: Routledge.

Wenger, E. (1998). *Communities of practice: Learning, meaning, and identity*. New York: Cambridge University Press.

Wertsch, J. (Ed.) (1985). *Culture, communication and cognition: Vygotskian perspectives*. Cambridge, UK: Cambridge University Press.

Williams, J. S., & Wake, G. D. (2004). Metaphors and cultural models afford communication repairs of breakdowns between mathematical discourses. In M. Holmes & A. Fuglestad (Eds.), *Proceedings of 28th annual conference of the International Group for the Psychology of Mathematics Education*. Volume 4, (pp. 409–416). Bergen, Norway: PME.

Yackel, E., & Cobb, P. (1996). Sociomathematical norms, argumentation and autonomy. *Journal for Research in Mathematics Education, 27*(4), 458–477.

CHAPTER 3

LANGUAGE IN MATHEMATICS TEACHING AND LEARNING

A Research Review

Mary J. Schleppegrell

ABSTRACT

My goal in this chapter is to identify key themes and major ideas in discussions of mathematics and language, focusing on central challenges and opportunities for work that would advance the field of mathematics education through a better understanding of the relationship between language and mathematics as it presents itself in mathematics classrooms. As a linguist, I bring the lens of an applied linguistics perspective to this project. The sources through which I identified relevant work reflect my disciplinary leanings and are not comprehensive of the literature in the field. I have tried to be representative in discussing work that has made contributions to our understanding, with a particular focus on illuminating the contributions of linguists and applied linguists so that their work is better understood in the mathematics education community. Understanding the ways researchers have seen the relationship between mathematics and language and identifying what is known about the features of the language of mathematics can help us formulate a research agenda that leads to greater understanding of the linguistic challenges of

Language and Mathematics Education, pages 73–112
Copyright © 2010 by Information Age Publishing
73

constructing mathematics knowledge and better preparation of teachers for engaging all students in learning mathematics.

OVERVIEW

The words "language and mathematics" can be thought of in two different ways: as referring to their relationship as systems of meaning-making and as referring to the role of language in the pedagogical context of mathematics classrooms. Both of these aspects of the relationship of language to mathematics are addressed in this review. Language has been a topic of discussion by mathematics educators for more than a generation. Stimulated by seminal work in the 1970s and 1980s (e.g., Halliday, 1978; Pimm, 1987), the notions that mathematics is itself "language-like" and that language issues are an important focus in classroom teaching have been consistent themes in mathematics education research. The idea that communication and discourse are integral to learning mathematics is now explicit in mathematics standards, and researchers have focused on the role of language and interaction in the mathematics classroom from a variety of perspectives that go by labels such as *constructivist, sociocultural,* and more recently, *semiotic.* This review discusses trends in research on language and mathematics, focusing in particular on recent work that recognizes the role of language in constructing knowledge and that has described the features of the mathematics register in comprehensive ways, recognizing its multi-semiotic nature (e.g., Lemke, 2003; O'Halloran, 2005).

Although in the past it has been suggested that mathematics draws less on language than other school subjects do, that view has now been discredited. In fact, it has become clear that language plays as important a role in mathematics learning as in learning other school subjects. Mathematics educators have gone beyond vocabulary and word-based thinking about language, as words alone do not suffice in identifying and describing the language challenges of mathematics (although in practice even some current research still focuses on language as meaning at the level of the word or phrase). The notions of mathematics as *discourse* and students as being apprenticed into particular ways of *doing mathematics* in particular discursive contexts are now gaining prominence in mathematics education research.

Every school subject is constructed in language, but the forms and patterns language takes vary from discipline to discipline. The language through which mathematics is constructed tends to be conceptually dense, interpersonally alienating, and highly structured textually in unfamiliar ways. A key challenge is that mathematics incorporates a symbolic language that developed out of natural language and also uses visual display to construct complex meanings. Students need to be able to work simultaneously with all aspects of this multi-semiotic system—natural language, the language of mathematics symbolism, and the visual semiotic constructed in the graphs, charts, and diagrams that are integral to mathematical reasoning. Recent research on language and mathematics using semiotic approaches and linguistic tools is providing insights into the ways these meaning-making systems interact in construing mathematical knowledge and is exploring the ways students are positioned by language in mathematics classrooms. This research offers possibilities for new understanding of the role of the teacher in supporting students' development of mathematical knowledge.

The first section of this review discusses different perspectives on the conceptual relationship between language and mathematics and describes semiotic approaches that recognize the importance of language in constructing mathematical knowledge. The second section focuses on language in the mathematics classroom, where researchers have described the dilemmas teachers face in moving between everyday and technical language in negotiating mathematics meaning in classrooms with diverse students. The third section describes studies that use functional linguistics perspectives to analyze language in mathematics classrooms using rich theoretical frameworks and drawing on rigorous analytic tools. The fourth section addresses the potential for teacher education to promote greater understanding of the role of language in mathematics education, and the final section provides a summary of recommendations for future research that emerge from this review.

LANGUAGE AND MATHEMATICS CONCEPTUALIZED

The relationship between language and mathematics was developed in a frequently cited seminal work by Pimm (1987), who used the metaphor that *mathematics is a language* and summarized a set of linguistic phenomena that characterize the language of mathematics. These included mathematical usage of ordinary words with different meanings (*difference, odd/even*), technical terms particular to mathematics, and different ways of interpreting words and phrases in mathematics. But Pimm also pointed out that mathematics is like language in that it is a system for making meaning. He drew educational implications from this understanding, suggesting that mathematics instruction needs to emphasize meaning and not form so that it is not treated as a kind of blind symbol manipulation that cuts students off from the real meaning of mathematics. Stimulated by the work of Pimm and others, the mathematics education community took up the notions of *discourse* and *communication* as central to mathematics learning. At the same time, individualist orientations that were proving inadequate for understanding mathematics teaching and learning stimulated interest in the notion of mathematics as social practice (Solomon, 1989).

Two perspectives on the relationship between language and mathematics can be characterized as *constructivist* and *socioculturalist*. While researchers define the approaches and position themselves relative to them in different ways, the constructivist view comes out of a Piagetian perspective, that everyone constructs internal representations, or mental structures, for him/herself. Cobb and Yackel (1998) characterize the constructivist perspective as interpretive, seeing knowledge as actively constructed by the student in interaction with the environment. Constructivists interpret the way students talk about mathematics to investigate students' development of mathematical knowledge (see also Laborde, Conroy, De Corte, Lee, & Pimm, 1990). Sociocultural perspectives (e.g., Adler, 1997) focus on discursive practices and the social engagement of students. They draw on Vygotskyan frameworks that stress the interaction between language and cognition and highlight the social dimension of learning and the role of communication and participation. In particular, Vygotsky's proposal, that all learning begins in interaction and that it is through social interaction through language that children develop the more formalized concepts of schooled subjects, has been very influential.

These two prominent views—on the one hand, that students construct mathematics knowledge for themselves as they develop the mental representations that are necessary for doing mathematics, and the other that students are enculturated into mathematics through social and discursive activity (and of course this simplifies both positions)—are currently being extended in approaches to research on mathematics education that call themselves *semiotic* (see, for example, the papers in Anderson, Sáenz-Ludlow, Zellweger, & Cifarelli, 2003 and Cobb, Yackel, & McClain, 2000). Semiotic approaches take different forms and draw on different epistemological orientations, but they recognize both language and mathematics as complex meaning-making systems. Semiotic approaches have in common the view that language is more than a tool for representation and communication. It is a tool for thinking and constructing knowledge. Many researchers who work from a semiotic perspective are also interested in interpersonal and developmental aspects of learning. Sfard (Sfard, Nesher, Streefland, Cobb, & Mason, 1998; Sfard, 2001), for example, suggests that "all our thinking is essentially discursive," (Sfard et al., 1998, p. 50) and puts forward what she calls a *communication approach* to cognition (Sfard, 2001). McNamara (2003) characterizes a semiotic approach as understanding that "[k]nowledge is actively constructed by the cognizing subject, not passively received from the environment... [and] [c]oming to know is an adaptive process that organises one's experiential world..." (p. 30). She suggests that it is necessary to adopt this understanding in order to see that language is not just a labeling process in mathematics, but that in fact language is the means for developing mathematical ideas.

Semiotic and information processing approaches are contrasted by Radford (2000), who analyzes students' thinking about algebra. From the information processing point of view, students are expected to extract meaning from the structure of the language as they manipulate symbols. But this point of view does not acknowledge the contribution of the students' experience and the social nature of the language and context for learning. Radford points out that students need to learn to reason in new ways in learning mathematics and that their prior knowledge and experience shape this in different ways. If we understand thinking as social practice in context, it is clear that students coming from different cultural contexts will be positioned in different ways to take up mathematics knowledge.

These differences can be acknowledged and made resources in mathematics classrooms if the focus is on meaning and if teachers are able to hear different perspectives. Sáenz-Ludlow (2003) argues that a semiotic approach can transcend arguments about whether it is the cognitive activity of the individual that is primary, with social interaction necessary but secondary, or whether knowledge is constructed as people interact. Her point is that cognitive activity and social interaction co-exist and co-act synergistically to support evolving understanding that involves both interpretation and the construction of mathematical meanings. She points out that when there is interpretation and construction, there will also be self-expression, again highlighting the interpersonal aspects. She suggests that "semiosis is a signifying process in which thought, language, and culture interweave to induce and sustain interpretation, construction, and expression of knowledge" (Sáenz-Ludlow, 2003, p. 260). She recommends "interpreting games" that enable the teacher to hear how students interpret mathematical meanings, using different wording to express the meanings in mathematical notation. Such interpretation, she suggests, enables movement from concrete contexts of doing mathematics to more abstract mathematical understanding. Students begin to understand the links among concepts through their own idiosyncratic symbols and formulations.

Linguistic tools, based on a social semiotic view of language, enable researchers to link expression with meaning in comprehensive and theoretically motivated ways so that language can be related to context in ways that show how meaning is made in the different activities of mathematics education. Sfard and Lavie (2005) point out that researchers get more reliable results when they analyze children's exact words, as analyzing actual utterances guards against interpretations of children's talk that are influenced by adult understanding. They use linguistic evidence to examine the way notions such as *comparison* are used by young children, analyzing how they apply the adjectives *bigger* and *smaller* or the adverbs *more* and *less* and show that young children's talk about numbers is not objectified in the way adults' is, and that the way the children talk about comparison shows that they are not using the same meaning systems as adults do. Sfard and Lavie point out that it is the use of language and engagement with concepts that provide the experiential contexts where children come to shape their meanings toward adult understanding.

Sfard and Lavie (2005) suggest that the path from everyday knowledge into the more specialized knowledge of formal mathematics can be thought of as the development of a specialized discourse that requires *objectification*. They illustrate how, in learning arithmetic, young children begin with ritualized participation in routines of counting and interacting with adults that then develop through discourse into more formalized knowledge. They recognize objectification when children begin using number words to signify entities that are not just concrete objects in the immediate context. For example, when children use number words as determiners (e.g., *three blocks*), they are not objectifying, whereas when they begin to use number words as nouns (e.g., *three is greater than four*), objectification is recognized.

The notion that close analysis of the forms of language itself is important for understanding the meanings that are being construed is formalized in the social semiotic approach of systemic functional linguistics (SFL), which provides a comprehensive framework for linking linguistic realization (the way the concepts are articulated in language) with the meaning that is thereby construed. SFL is based on the notion that meaning, and therefore also language, varies according to social and cultural context. It suggests that the choice of the grammatical form (*choice* in terms not of conscious selection but of unconscious use) varies according to the contexts in which speakers interact, and that analyzing language choices can reveal important differences in how content, role relationships, and information flow are constructed by different speakers in different contexts. From this perspective, analysis of the actual language used in mathematics teaching and learning reveals the knowledge that teachers make available in mathematics classrooms and how students take up new knowledge in the context of actually doing mathematics, as well as the way students are positioned and engaged as they learn (examples to illustrate this are presented below).

SFL provides an elaborated grammar of English that links meaning with the grammatical forms through which meaning is made (Halliday & Matthiessen, 2004) and enables exploration of the patterns of language through which meaning is constructed in interaction and in pedagogical texts. This social semiotic approach was introduced to mathematics educators in Halliday's early essay on the mathematics register (Halliday, 1978), often cited by researchers. Without a comprehensive understanding of the linguistic framework, however, researchers sometimes interpret the mathematics register as merely a set of words and phrases that are particular to mathematics.

Halliday defines *register* as

> a set of meanings that is appropriate to a particular function of language, together with the words and structures which express these meanings. We can refer to a 'mathematics register', in the sense of the meanings that belong to the language of mathematics (the mathematical use of natural language, that is: not mathematics itself), and that a language must express if it is being used for mathematical purposes. (1978, p. 195)

He stressed that this is not just a question of learning new words, but also of learning new "styles of meaning and modes of argument... and of combining existing elements into new combinations" (p. 195–196). Halliday points out that counting, measuring, and other "everyday" ways of doing mathematics draw on "everyday" language, but that the kind of mathematics that students need to develop through schooling uses language in new ways to serve new functions. The notion of a mathematical register helps us see in language how mathematical knowledge is different from knowledge in other academic subjects and recognize the ways students need to be able to use language to effectively participate in the ways of knowing that are particular to mathematics.

Learning the language of a new discipline is part of learning the new discipline; the learning is not separate from the development of the language that constructs the new knowledge. As with all school learning, a key challenge in mathematics teaching is to help students move between everyday, informal ways of construing knowledge and the technical and academic ways that are necessary for advanced learning. Since students come to school with everyday language with which they have constructed their knowledge of the world, the school can build on that knowledge and language to move students toward new, more scientific and technical understandings through consciousness about and attention to the linguistic challenges that accompany the conceptual challenges of learning.

The work of Jay Lemke has been seminal in drawing on SFL theory to focus on the way mathematical knowledge is constructed. Lemke (2003) shows that the mathematics symbolic language developed out of natural language so that the two systems are integrally related, as "[t]he history of mathematical speaking and writing is a history of the gradual extension of the semantic reach of natural language into new domains of meaning" (p. 217). Going beyond Pimm's metaphor that "mathematics is a language," Lemke shows in nuanced ways just how language and mathematics are related, describing how, as mathematics developed historically, it drew on and developed out of natural language, creating the mathematics symbolism and drawing on visual display in distinctive ways.

Lemke has been a major contributor to the evolution of a social semiotic view of mathematics, and in addition to his description of the evolution of mathematics language, he has made two other key contributions: (1) the development of the notion of *thematic patterns*, based on his analysis of language in science classrooms (1990), and (2) the description of the interaction of the multiple semiotic systems that work together in mathematics (2003). Thematic patterns are the patterns of meaning built up by teachers and students as particular topics are developed in mathematics classrooms. Lemke (1990) calls mastery of thematic patterns the most essential element in learning. He demonstrates that when students use a different thematic pattern than the teacher does in talking about a topic, they are also thinking about it in different ways. The different semantic combinations have different meanings, and in seeing the meanings the students construe, we can see what they are understanding and taking up from the teacher and where their understanding does not fit with the official discourse of mathematics. Lemke points out that "[i]n teaching... any subject, we do not want students to simply parrot back the *words*. We want them to be able to construct the essential *meaning*s in their own words, and in slightly different words as the situation may require. Fixed words are useless. Wordings must change flexibly to meet the needs of the argument, problem, use, or application of the moment. But they must express the same essential *meaning*s if they are to be scientifically acceptable and, in most cases, practically useful. This is what we mean when we say we want students to "understand concepts" (1990, p. 91).

Lemke shows that we can explore the concepts students are developing by analyzing the thematic patterns that they use. Chapman (1995) applies these insights in her analysis of an algebra lesson, showing how a teacher builds up the notion of a *constant difference* in finding the slope of a line by starting with a visual representation, a pattern of dots. Chapman traces the thematic patterns in the teacher's development of the notion that *a pattern of dots generates numbers* that *lead to a rule*. The procedure for finding the rule involves completing an *x + y table* and looking for the *common difference*. Over the evolution of the teacher's explanation, the formulation *the numbers* is reformulated as *the table*, the *difference pattern* is reformulated as the *constant*, and the *constant* is said to *lead to* the *linear* pattern. Students need to be able to follow the semantic shifts that occur as an explanation like this evolves. Tracing the thematic patterns shows the conceptual complexity and the challenges students face in making the connections in the way the mathematics understanding requires. Analysis of thematic patterns offers a methodology for identifying key concepts, seeing how they are presented to students and how students take them up, providing a powerful tool for investigating students' learning.

SFL theory also enables researchers to take into account the multi-semiotic nature of mathematics. Lemke (2003) points out that mathematics discourse enables us to present "meanings that natural language has trouble articulating" (p. 214) through the interaction of natural language, mathematics symbolism, and visual display. Spatial-visual proportions and continuous covariation of discrete operations on continuous variables can be represented in mathematics discourse where natural language, mathematics symbolism, and visual representations "form a single unified system for meaning-making" (Lemke, 2003, p. 215). Lemke points out that discrete (typological) meanings are the domain of natural language, where we focus on discrete processes and things, while continuous (topological) meanings are the domain of mathematics, where fractions and complex ratios are easier to construe through mathematics symbolism and in visual display than in natural language. But it is the three systems working together that make up mathematical reasoning.

SFL theory provides a means of understanding how semiotic choices work together so that students can be made aware of the different strategies used for construing meaning in language, symbolism, and visual images. In his analysis of a physics classroom discussion, Lemke (2003) shows that

> it would not be possible to get a complete and correct meaning just from the verbal language in the activity, nor just from the mathematical expressions written and calculations performed, nor just from the visual diagrams, overheads, and chalkboard cues, nor just from the gestures and motor actions of the participants. It is only by cross-referring and integrating these thematically, by operating with them as if they were all component resources of a single semiotic system, that meanings actually get effectively made and shared in real life. (p. 229)

He also suggests that "[t]oo much opportunity for gaining mathematical understanding and intuition, too much practice at learning how to use mathematical meaning in real situations, is lost if mathematics is not taught, particularly at the introductory level, as a co-equal partner with language and visual representation in the analysis of natural and social phenomena" (2003, p. 231).

The linguistic features of the mathematics register have recently been more fully elaborated by O'Halloran (2005) in a comprehensive grammatical description of the register features of mathematical discourse. She describes the roles of each of the semiotic systems, pointing out that natural language plays a major role in contextualizing problems, explaining the sequence of activities that need to be undertaken, and discussing the implications of the results:

> Language is often used to introduce, contextualize and describe the mathematics problem. The next step is typically the visualization of the problem in graphical or diagrammatic form. Finally the problem is solved using mathematical symbolism through a variety of approaches which include the rec-

ognition of patterns, the use of analogy, an examination of different cases, working backwards from a solution to arrive at the original data, establishing sub-goals for complex problems, indirect reasoning in the form of proof by contradiction, mathematical induction ... and mathematical deduction using previously established results. (p. 94)

The mathematics symbolism developed out of and depends on natural language. "[M]athematical symbolism developed as a semiotic resource with a grammar through which meaning is *unambiguously* encoded in ways which involve *maximal economy* and *condensation*" (O'Halloran, 2005, p. 97). The symbolism enables mathematical meanings to be presented in exact ways, but in doing so it condenses meaning into structures that are not found in natural language, and so requires specialized knowledge about how this semiotic resource works. At the same time, the condensation and economy are functional for making the kinds of meanings that mathematics enables. O'Halloran points out that "mathematics is seen to deal with a limited semantic field in limited ways, but in doing so has the potential to solve problems which would be impossible to solve using other semiotic resources" (p. 24).

Building on Halliday's definition, O'Halloran shows how the mathematics register can be systematically analyzed and described with a theoretically grounded means of linking form and meaning. SFL theory recognizes three kinds of meanings that are simultaneously realized in all mathematics discourse: *ideational* meaning that is both *experiential* and *logical, interpersonal* meaning that positions interlocutors in particular ways, and *textual* meaning that organizes and presents information. Each of these kinds of meaning can be linked with the linguistic expressions that construe it. To simplify somewhat, *experiential* meaning is realized in verbs, nouns, prepositional phrases and adverbs; *logical* meaning is realized in conjunctions and connectors; *interpersonal* meaning is realized in clause mood (whether statements, questions, or commands) and modality (degrees of *likelihood, obligation, usuality,* etc.); and *textual* meaning is realized in the way information is organized and presented.

The central element of the clause, the basic unit of language, is the *process,* constructed in the verb or verb phrase. Each process may have associated *participants* (typically constructed in noun phrases) and *circumstances* (typically constructed in adverbs and prepositional phrases). The mathematics register draws on only a subset of the elements of the linguistic systems that characterize natural language to construe experiential meaning. SFL distinguishes six process types that occur in natural language (*material, mental, behavioral, verbal, relational,* and *existential*). O'Halloran demonstrates that mathematical discourse, mainly concerned with describing and manipulating relations, uses few *material, mental, behavioral,* or *verbal* processes; instead, *relational* processes (processes of *being* and *having*), and *existential* processes

(processes of *existing;* e.g., *Given. . . .*) dominate. O'Halloran identifies a seventh process type, the *operative* process (processes of *addition, subtraction,* and other calculations), as an innovation of mathematics discourse. Operative processes and their participants enable the mathematical symbolism to reconfigure structure configurations in solving problems; so, for example, $s(t) = -16t^2 + 80t$ is a combination of "components" ($s(t)$, 16, t, 80, +) in "expressions" ($-16t^2$, $80t$) that have been configured into a "clause" where, for example, the expression $-16t^2$ is a reconfiguration of operative processes and participants ($-16 \times t \times t$) (examples from O'Halloran, 2000). O'Halloran (2005) points out that the mathematical symbolism "is functionally organized to fulfill the goals of mathematics: to order, to model situations, to present patterns, to solve problems and to make predictions" (p. 108). The *participants* in mathematical processes typically are numbers and variables that function as general representations rather than specific entities. The *circumstances* are only those that describe relationships and not the full range of circumstances used in natural language. O'Halloran characterizes the experiential meanings typical of the mathematics register as an "expanded realm of meaning within a restricted experiential field" (p. 110). She points out that spatial and positional notation also realize experiential meaning in forms not found in natural language, and that all of these features make maximal structural condensation possible.

In terms of interpersonal meaning, O'Halloran shows that "[t]he objective and factual appearance of mathematics results from a combination of the restricted selections in the fields of experiential and interpersonal meaning in the mathematical symbolism . . . the textual strategies of condensation through which meaning is efficiently encoded, and the emphasis directed towards logical meaning" (2005, pp. 113–114). The logical meaning depends on knowledge and principles that are often not made explicit in the mathematics text, adding to the challenges for students. The textual organization of mathematics is "highly formalized in order to facilitate the economical encoding of relations which permits immediate engagement with the experiential and logical meaning of the text" (O'Halloran, 2005, p. 121). Again, here the use of spatiality is a key element, as the reading path is not necessarily linear, and mathematical discourse permits more ellipses (expressions that are left unstated but assumed) than natural language. This textual organization also permits maximum condensation. All of these resources, then, meet the aims of encoding patterns of relations economically and exactly, and permit reconfiguration of elements as needed as problems are solved. The complexities of the grammatical descriptions O'Halloran develops are too extensive to review further here, but as the research based on this work is described below, the approach will be better elucidated and some of the constructs elaborated.

O'Halloran (2005), like Lemke (1990), recommends that the nature of the language itself become a point of discussion with students and that they engage in analysis of the ways the different elements of mathematical discourse interact. She suggests that through this approach,

> students may come to understand that language is a tool used to create order, and that...content is only one aspect of the order.... Equally important are the social relations which are enacted, the logical reasoning which takes place and the ways in which the message is organized and delivered. In addition,...students can appreciate that there are culturally specific ways in which language is used in different contexts. The understanding of those ways, and the interests served by such language selections, opens the way for a critical engagement with texts. (p. 200)

Summary

Mathematics education researchers have sometimes constructed a dichotomy between the *cognitive* and the *social*, with different researchers staking out territory in one or the other of these areas. A social semiotic perspective allows a synthesis of these views, where cognition is seen as situated activity that is socially constructed and enabled through language and interaction. What the social semiotics of systemic functional linguistics adds to the semiotic approaches more broadly is a language for talking about language that enables analysis of the meanings construed by the three semiotic systems used in mathematical discourse (natural language, mathematics symbolism, and visual display) in interaction with each other. The social semiotic perspective based in systemic functional linguistics recognizes that differences in wording and meaning have implications for the learning that is taking place and the forms of consciousness about mathematics that students develop. Work drawing on this perspective can help us understand the role that language and classroom interaction play in providing students with access to mathematics and can help us see where students in some classrooms may not be getting such access. SFL lets us look at the processes of teaching, learning, and doing mathematics by identifying the meanings that are made in the texts and interaction. Language is seen as *construing* meaning; actually making the meaning available through the realization in spoken or written language. The meanings that are construed are related to the contexts of use, and these are recognized at both local (context of situation) and more global (context of culture) levels. Using a lens that focuses on one or the other level, or both, provides insights into the meanings that are made.

Seeing language as constitutive of meaning allows us to analyze the language produced by students and teachers in classrooms, or the language of the textbooks and other explanations students work with, to recognize differences in the ways language is used to construct the content knowledge of mathematics and the ways it is used to contextualize that knowledge and facilitate students' interaction with it. As we will see, a major tension teachers face is in finding middle ground between the knowledge and language students bring and the knowledge and language they need to develop in mathematics classrooms, and a means of talking explicitly about the language through which mathematics is constructed offers a way of responding to this tension. As is explored below, researchers have used SFL to show how students in classrooms of different kinds experience mathematics differently and are offered differential opportunities to get access to this powerful discourse. These insights enable us to recognize differences in the ways mathematics is construed in schools with students from different linguistic, social class, or ethnic backgrounds and to link the construals with opportunities to learn mathematics, providing us with ways to intervene and notions of what can be improved in mathematics education. This review will return to these points after considering the classroom pedagogical issues that research in mathematics education has raised and discussed.

LANGUAGE IN MATHEMATICS CLASSROOMS: DILEMMAS AND TENSIONS

Studies of language in mathematics classrooms raise concerns that connect with the theoretical considerations discussed above. How is mathematics different from other school subjects? How can talk about language support students' mathematics learning? How explicit and precise should a teacher be in presenting mathematics concepts and how much everyday language and technical language should be used? How can teachers of diverse learners use language effectively in teaching mathematics? How does language affect assessment of students' mathematical knowledge? Research on these questions is reviewed here.

The two key subjects in school learning are language arts and mathematics, and some work on mathematics education has promoted strategies from language arts such as process writing and a more narrative approach as a way of making mathematics learning less formal or less alienating. Such practices have been critiqued, however (see discussion in Schleppegrell, 2007). Mathematics is different from language arts in the way it develops knowledge through relatively precise language and formal models (Bernstein, 1996), drawing on language that is different from the language that is typically used in language arts. Solomon and O'Neill (1998), for example, point out that mathematical argument achieves cohesion through logical rather than temporal order, different from narrative. They point out that "[w]hile it is right to value children's own knowledge and experience, it is necessary... to teach them about different literacy practices which have different functions... and cannot be generated simply from everyday practices which have very different functions" (p. 219). They describe the genres through which mathematical meanings are constituted and suggest that those genres should be a focus of instruction (see also Marks & Mousley's (1990) critique of process approaches to writing in the context of mathematics teaching). Attempts to make mathematics relate directly to students' experience and to eliminate or downplay the technicality of mathematics may also be problematic, since the technicality is functional for making the kinds of meanings that are relevant to constructing knowledge in mathematics. As Pimm (1987) points out, although many people talk about how precise mathematics language is, the precision is in the way language is used. The technical language has to be practiced and developed along with the mathematics concepts.

As work on classroom discourse has illuminated classroom interactional practices, it has become clear that a focus on discourse is not just a question of stimulating talk in classrooms. While most mathematics educators agree that it is important for students to use language to explore mathematical issues, the kind of exploration that is needed has to be particular to the kind of knowledge that mathematics constructs. Silver and Smith (1996) point out that classroom talk needs to ensure that the mathematics does not get lost, and that the discourse centers on "worthwhile tasks that engage students in thinking and reasoning about important mathematical ideas" (p. 24). If the discourse centers only on giving the right answer or on procedural issues, it does not model mathematical thinking and reasoning. Teachers co-construct mathematical explanations more effectively with some students than with others, providing different opportunities for students to develop the mathematics register (Forman, McCormick, & Donato, 1997). To effectively interact in ways that support mathematical development, teachers need to recognize what a student is trying to say and improve the student's ability to articulate it rather than seeing language merely as a way to assess students' knowledge (Laborde et al., 1990).

The notion of precision in the representation of concepts and clarity about what is being communicated has been a prominent theme in work on language and mathematics and has led to much discussion of the dilemmas teachers face in reconciling precision and clarity with the messy task of initiating students into the technicalities of mathematics discourse. Related to this is the role of everyday language in mathematics learning. Some mathematics concepts can be articulated in everyday language, but some require using language in new ways. For example, MacGregor (2002) found that students need to develop new ways of using language in order to structure mathematical concepts in the precise ways that are required for participation in advanced mathematics contexts. She illustrates this with an example of comparison as it was construed by the Australian students and pre-service teachers that she studied. A student explains, *if Tina has twice as much money as George, then George has twice as less than Tina* (p. 83). The student is using *twice as less* instead of *half as much,* but such usage does not construct the concepts of difference and comparison with the precision that is needed to enable further development of these concepts in higher mathematics. MacGregor suggests that "the grammatical form of a comparative expression plays a part in determining the form of the associated concept" (p. 80) and that "grammatical form ... can reveal that a concept is vague, under-developed, unstable or incorrect" (p. 80). Teachers need to recognize when students are using such expressions and understand their value in the colloquial language of students' home communities, but also help students adopt the mathematical discourse that will enable them to participate in mathematics in the formal context of schooling. MacGregor (2002) found that "secondary students who described a relation between numbers in an informal, unclear or immature way were unable to relate it to a mathematical operation" (p. 79). There are many such forms that are common in everyday situations, and such everyday ways of using language are not adequate for constructing the precise and technical meanings that mathematics requires.

In the pedagogical context, the tensions related to the notions of everyday versus technical language and explicitness versus implicitness are often talked about in terms of the language of "dilemmas," following Lampert (1985). Adler, for example, in the complex multilingual context of South African classrooms, has discussed three key dilemmas: the *dilemma of mediation* (Adler, 1997, 1999), the *dilemma of transparency* (Adler, 1998, 1999), and the *dilemma of code-switching* (Adler, 1998). Adler's (1997) *dilemma of mediation* focuses attention on the need for teachers to listen to and validate the perspectives learners bring, while at the same time moving them from their own formulations of concepts into the formalized discourse of mathematics. In her case study of a South African classroom where the teacher uses both student grouping and teacher–student interaction, Adler shows how enabling students to work together, without monitoring by the teacher, enables a participatory classroom culture, but that the teacher's intervention and mediation "is essential to improving the substance of communication about mathematics and the development of scientific concepts" (1997, p. 255). Adler suggests that the dilemma is "in shaping informal, expressive and sometimes incomplete and confusing language, while aiming toward the abstract and formal language of mathematics," pointing out that "a participatory-inquiry approach, and the possibilities it offers for learner activity and pupil-pupil interaction, can inadvertently constrain mediation of mathematical activity and access to mathematical concepts" (p. 236).

The teacher's oral language is a key means of linking between visual and symbolic representations, making the spoken language very powerful in classroom learning and suggesting that the way language and mathematics interact can become an explicit focus of attention in classrooms so that knowledge can be presented to students in explicit ways (Veel, 1999). But the notion of explicit teaching, and making the language transparent in its meaning, is not a straightforward issue. Adler (1998) demonstrates that explicit mathematics language teaching, focused on instructions and explanations, helped all learners, but the teachers she interviewed felt uncomfortable with all the talking they were doing. They felt that their attempts to make everything clear were sometimes distracting from the development of the mathematics concepts. This is an issue Adler (1998) calls the *dilemma of transparency*, whether teachers should step in to clarify concepts or not. Gorgorió and Planas (2001) report on classrooms where teachers working with immigrant students in Spain found simplified language of little help in communicating mathematical ideas. A simple change in vocabulary was seldom effective in clarifying concepts, and simplified forms of language sometimes obscured mathematics knowledge rather than clarifying it, although in a search for transparency, teachers still often tended to simplify. This research showed that problem-solving activities in linguistically homogeneous groups minimized some of the linguistic difficulties, and Gorgorió and Planas call for more research "to clarify how mathematical language can be taught and to investigate the relationships between the 'language of the mathematics class', mathematical language, and the process of constructing mathematical knowledge" (2001, p. 30).

These issues are especially crucial for teaching and learning in multilingual and multicultural classrooms. Schooled language is a challenge for all students, but there are particular groups of students for whom the challenges are greater than for others. English language learners and speakers of nonstandard varieties of English are two of these groups. Adler's *dilemma of code-switching*, referred to above, addresses some issues that come up in multilingual classrooms where teacher and students share a common language in addition to English. *Code-switching* refers to the practice of moving between languages in seamless ways, a practice common in bilingual and multilingual settings. Setati and Adler (2001) describe code-switching in a South African multilingual context where mathematics is taught in English, but the students and teachers also speak Setswana. The students in these classrooms hear English only at school, and although the teachers understand that talking together is a way of thinking together, using the students' own language is problematic when that language has not developed a mathematics register that engages with the kind of mathematics taught at school. This is the case for many languages in the world that are unwritten or not developed for higher education, such as Setswana, and also for nonstandard dialects of English such as African American vernacular. It does not denigrate students to say that their home language does not include an advanced mathematics register. For historical reasons not every language or variety of language has evolved the specialized ways of meaning that construct the mathematics students are expected to learn in school. As Lemke and O'Halloran illustrate, mathematical discourse evolved over time out of natural language in the specialized contexts of mathematics use in specific cultural and linguistic contexts. Not every language variety has a register that has developed to do this kind of mathematics. Every language and dialect has the potential to develop such a mathematics register, but there is little social context for or political interest in investing in such development today for languages that are not already used for the kind of mathematics discourse done at school.

One way to address this is by recognizing the different registers that operate simultaneously in classrooms, so that the registers students bring can serve as resources in the development of new registers. In the South African classrooms she studies, Setati (2005) distinguishes four registers (she calls them *discourses*), referring to when the language is about procedural steps to solve a problem, when the focus is on reasons for calculating in particular ways, when the language builds the context of word problems, and when the language is used to regulate students. The teacher moves between English and Setswana for these different purposes. Setati points out that English is the language of assessment and so provides access to the social goods that come through learning mathematics but at the same time perpetuates the hegemony of English in South Africa. Not providing access to English risks marginalization of the students at the same time that the teacher needs to use Setswana to manage much of what is going on. Setati suggests that teachers think about the different opportunities the registers offer for responding to and validating the language the students bring to the classroom on the one hand, while seizing opportunities to move the students in the direction of the more technical mathematics register when constructing mathematical knowledge. Working in contexts where students speak dialects or languages that do not have mathematics registers that can be used in classroom learning is a challenge, and research on this issue that recognizes the role of register and that can distinguish the use of different registers in the classroom is needed.

The role of the technical language is an issue recognized by research on English Language Learners (ELLs) in U.S. contexts as well. In bilingual settings it is important that teachers know the mathematics and can use the technical register in both languages where two languages are used and both have a mathematics register (e.g., English/Spanish contexts). Khisty and Viego (1999) suggest using technical language with Latino bilingual students and focusing on reasoning rather than correct answers in responding to students. Moschkovich (1999, 2000, 2002) analyzes teacher–student interaction in classroom episodes with Latino students in California. She shows how these students use gesture and other modalities in bilingual conversations to make meaning and participate in mathematical discourses. She points out that manipulatives and pictures cannot be used without language and grapples with the issue of the movement between informal and technical language, suggesting "revoicing" students' contributions to make them more mathematical and not correcting students' language errors (Moschkovich, 1999). She points out that teachers and learners are not always talking about the same thing and critiques the typical approaches offered through generic ESL strategies because they have no mathematics content or guidance on how to concentrate on the mathematics content of discussions.

Lager (2006) also illustrates that generic ESL strategies are not enough. He points out that teachers know they need to create and use manipulatives, form cooperative groups, and increase their content knowledge, but they do not have rich knowledge about the role of language in facilitating these strategies or in making mathematics comprehensible and accessible to students. He reports on a study of ELLs and non-ELLs responding to a set of middle school algebra items about a linear pattern and shows how just one misunderstanding can lead to logical but incorrect solutions that then affect each subsequent item. Lager points out that "modeling problem situations requires translating from everyday language to algebraic expression...including the reorganization and reinterpretation of problem information" (p. 167).

It is also not practical to think that students can learn the language they need outside of mathematics classrooms and so be prepared, for example, by ESL teachers who themselves are not mathematics teachers, to deal with the language of mathematics. Barwell (2003, 2005a, 2005b) analyzed nine- and ten-year-old ELLs interacting in mainstream classrooms in the UK as they wrote word problems, focusing on what they attended to in completing the task. He found that these students paid regular attention to mathematical structure, to the shape of the genre, and to written aspects of the work (spelling, punctuation, tenses), and pointed out that for the students, the content and language were not separate. Barwell calls for "a more explicitly reflexive model of the relationship between content, language, and learning" (2005a, p. 206) that sees language and mathematics as jointly constructed and not separate. These studies reinforce the notion that the movement between the language that students bring and the new ways of using language they need to develop depends on teachers' abilities to recognize how best to construct mathematical meanings in language.

English language learners are a heterogeneous group and they do not all face the same challenges in school learning. One useful distinction to make is between those who have developed literacy and worked at grade level in their first language and those whose education has been interrupted or who have never had an opportunity to develop literacy in their first languages. Students with literacy and grade level knowledge in their first languages will be able to move more quickly into the same content knowledge in English, as they will already have developed understanding of mathematics concepts in that language and will be aware of the mathematics register in their first language. In contrast, students whose only opportunities to learn are in the classroom and whose lives outside of school do not offer them experience in talking about mathematics or engaging with mathematical meaning will have more difficulty and will need more experience in order to develop mathematics knowledge and control of mathematics meaning-making. Such students share many challenges with speakers of nonstandard dialects who do not encounter academic English outside of school.

Cultural differences related to language use also affect expectations about student-teacher interaction. For example, in the classrooms Gorgorió and Planas (2001) report on, students have difficulty communicating in Catalan, the language of instruction, even when they "know" the language, because of cultural constraints on telling the teacher when they do not understand and because of other different communication norms. The teachers they studied also did not recognize that students could continue to have difficulty understanding mathematical discourse even when they seemed fluent in Catalan for other purposes.

Notions of universality and cognitive invariance across cultures are currently being critiqued, with accompanying interest in descriptions of the processes of learning and how they might differ across cultures (Sfard, 2001). Students may draw on different epistemologies and ways of knowing and doing mathematics that emerge from cultural differences (Selin & D'Ambrosio, 2000). Nunes, Schliemann, and Carraher (1993), for example, show how mathematics is used by fishermen, farmers, and carpenters in Brazil to solve problems informally and argue for a more realistic mathematics in the classroom (see also Presmeg, 1998). But even when they build on everyday mathematical practices, classroom activities also need to offer opportunities for learning academic mathematics. Cultural and linguistic differences are potential resources in the classroom when teachers can recognize different voices and differences in views and make mathematical knowledge an object of reflection in ways that enable more students to participate in talking about and learning mathematics (Zack & Graves, 2001).

The issue of authenticity is often raised in the context of discussion about mathematics word problems. Word problems are contrived by teachers and curriculum writers and do not draw on everyday knowledge, even when they are intended to link to authentic contexts. We cannot simplistically use what seem to be everyday contexts and expect that doing so will help students learn the mathematics they need to succeed at school. In fact, making it appear that mathematics draws on everyday knowledge in situations where in fact it does not may even make the knowledge less accessible to struggling students. As we have seen, mathematics is a technical discourse, and while teachers need to connect to students and what they bring in understanding to the classroom, they also need to move students into the disciplinary ways of talking about and doing mathematics that will enable them to participate in advanced contexts of mathematics learning.

Researchers have investigated how the language of mathematics word problems influences children's comprehension and ability to solve the problems, but the difficulties are typically located in the situations and contexts that the problems present (e.g., Staub & Reusser, 1995). Gerofsky (1996) has described the structure of word problems from a narrative/genre perspective, but we need to know more about the variables in word problems that affect students' comprehension, using frameworks that can link wording and meaning with students' difficulties. Some research related to ELLs' performance in mathematics has focused on the wording in high stakes exams. Abedi and Lord (2001) analyzed the effect of modifying the language of released mathematics items from the National Assessment of Educational Progress (NAEP) 1992 to make meaning more accessible to struggling learners. They changed the language of word problems in several ways: shortening nominal expressions, making conditional relationships more explicit, changing complex question phrases to simple question words and passive voice to active, and replacing less familiar or less frequently used non-mathematics vocabulary with more common terms. They then interviewed eighth grade students who worked both the original and modified forms of the problems, asking them whether anything was confusing or easy about the problems and whether they would choose first to do the original or revised problem. Most students chose the revised versions and performed better with those versions. Low-performing mathematics students benefited more from the revisions than those in higher mathematics and algebra, ELLs benefited more than proficient speakers of English, and low socioeconomic status students benefited more than others. Brown (2005) has also contributed to this agenda, investigating literacy-based performance assessments by third grade students in Maryland.

This is promising work that can contribute to the development of materials that provide better support for students who are moving from the everyday into the more technical language, and with a meaning-based grammar like that of SFL, we could engage in research that investigates variables such as how relationships between known and the unknown quantities are expressed, how explicit they are, what particular wording is chosen, and the features of distractors in multiple choice questions. Multiple choice questions comprise up to 70% or more of mathematics assessments (Veel, 1999), and we need to better understand to what extent mathematics questions test students' more general language skills as much as their understanding of mathematics. Such research could also help educators be clear about what we want students to learn and enhance the way standards are presented with more information about the language challenges and how teachers can address them.

Addressing the Dilemmas

The dilemmas discussed above are persistent issues that are central to mathematics teaching and learning. Students' mathematical ideas are shaped in the interactions they participate in, requiring that teachers be responsive to what emerges as students engage in mathematical work, while at the same time being accountable to the instructional agenda. Herbst (2003) discusses this issue as a set of *tensions* that teachers need to be aware of concerning "the direction of students' activity, the representation of mathematical objects, and the elicitation of the conceptual actions that students need to invest" (p. 198). As students do new tasks, they are developing new knowledge through the task, and teachers have to pay attention to the final product or goal while taking account of students' conjectures and arguments on the way. Teachers also face tensions in how to represent mathematical objects, as the precision that is relevant to the mathematics is important but so is some imprecision and vagueness in both teacher's and students' talk that allows students opportunities to think mathematically. Herbst's conceptualization of these dilemmas of mediation and transparency as tensions where no resolution is expected provides a practical way of thinking about the teaching process.

Teachers need to foreground everyday language in some contexts and technical language in others, and this has to do with the stage of development of the students' knowledge as well as the task that is currently being undertaken. In their conclusion to a special journal issue on language in mathematics education, Barwell, Leung, Morgan, and Street (2005) argue that "fuzziness, ambiguity, multiplicity of meaning and exploratory discussion in everyday language should be recognized, not as failure to achieve a truly mathematical degree of precision, but as essential to making mathematical meanings and to learning mathematical concepts" (p. 144). They recognize the tension between precision and exploration and ambiguity and suggest that teachers need more explicit awareness of "the variety of forms mathematical communication may take, as well as a need for resources to support them in working with learners to develop a fuller understanding of the nature and role of mathematical language" (p. 145).

Raising questions about the language teachers use to talk about mathematics leads us toward study of classroom language with an eye to finding new ways of presenting mathematics knowledge in language at different levels, in the contexts of different topics, and in different social contexts. With knowledge about the linguistic features of the mathematics register, teachers are better equipped to make shifts between everyday and technical language in systematic and principled ways. For multicultural classrooms, a way of talking about mathematics that allows students to share perspectives and talk about their understanding offers an approach to diversity that can respect different perspectives but also provide a common way of negotiating understanding. We know that mathematics learning benefits from scaffolding through social interaction with a more expert interlocutor, and studies that describe effective interaction of this type are needed at all levels and in all instructional contexts.

RESEARCH FROM A FUNCTIONAL LINGUISTICS PERSPECTIVE

This review has presented some theoretical perspectives that researchers have brought to the study of language and mathematics and has identified classroom issues that relate to language use in mathematics classrooms. Two key issues that a linguistic perspective can illuminate are highlighted here: the issue of how best to work with the multi-semiotic nature of the mathematics discourse itself and the issue of movement between everyday and formal ways of talking about mathematics. This section describes studies that illustrate how the linguistic tools and perspectives of SFL can help us understand and recognize how mathematics is construed in language and how students are positioned through language in mathematics classrooms.

Christie (1991, 2002) makes a useful distinction between the *content* or *instructional register* and the *regulative* or *pedagogical register* that projects the content. The technical mathematics discourse (the *content register*) is always embedded in and projected by the facilitative language of the classroom and text, the *pedagogical register*. The pedagogical register is the vehicle through which the content register is made available. Recognizing these registers in interaction with each other offers ways of negotiating the need to maintain an accurate representation of mathematics at the same time the language through which that knowledge is presented adapts to particular levels and contexts.

The mathematics content register itself has several dimensions. Research can explore the concepts that are presented, the logic of a text, the way a mathematics discourse positions students, and the textual organization of mathematics. In addition, research can explore how the mathematics content register is projected through the pedagogical register in different ways by teachers in different instructional contexts, with different kinds of learners, and at different stages and topics of mathematics instruction. With the linguistic tools that are now available, we are poised to be able to describe in greater detail, and with firm theoretical grounding, the ways meaning is made both in regulating classroom activity and in presenting the mathematics content, analyzing movement between everyday and technical ways of construing mathematics. In addition, recognition of the multi-semiotic nature of mathematics and the need to construct mathematics through movement among natural language, the mathematics symbolism, and visual representations opens new territory for research in the content register. Little has been done yet to explore how these systems work together, and research that explores this interaction in practice is needed. The studies described below provide some ways researchers have begun exploring these issues.

The Nature of the Knowledge Being Constructed in Mathematics Classrooms: The Content Register

Veel (1999) uses the SFL grammar to explore the features of the natural language used in mathematics and identifies distinctive ways that the elements of the clause appear in mathematics discourse. He points out that two types of relational processes, *attributive* and *identifying*, are pervasive in mathematics, and although they are often constructed by the same verb (forms of *is*), they present very different kinds of meaning. Attributive clauses classify objects and events, while identifying clauses introduce technical terms. An attributive process constructs information about membership in a class or part–whole relationship, as in: *A square is a quadrilateral* or *Three and four are factors of twelve*. An identifying process, on the other hand, constructs relationships of identity and equality, as in *A prime number is a number which can only be divided by one and itself* or *The mean, or average, score is the sum of the scores divided by the number of scores* (examples from Veel, 1999, p. 195). Identifying processes are particularly important because they can provide a bridge for students between technical and less technical construals of mathematics knowledge by enabling two formulations to be presented as equivalent (e.g., *Sides of the triangle that are in the same positions **are** <u>corresponding</u> sides of the triangles*). Relational processes are also a feature of the multiple choice questions that are often used to assess students' mathematics knowledge on standardized tests, as they ask *which of the following is correct/true/the best way*, etc. (Veel, 1999).

Clauses with the verbs *be* and *have* and other related verbs (*means, equals,* etc.) are challenging in their grammatical features. Students with first languages other than English may be accustomed to constructing relationships of attribution and identity in different ways than English. In Spanish, for example, the verb *is* has two different forms to construe the different meaning relationships. Veel is also interested in differences between the ways student use mathematical language and teacher/textbook use of mathematical language. He compares the lexical density (density of concepts) and the ratio of relational processes to non-relational processes in teachers', textbooks', and students' language and shows that there are gaps in the different formulations that are meaningful in terms of the knowledge constructed.

O'Halloran (2000) discusses the limitations of research on mathematics language that "largely centered around vocabulary, symbolism and isolated examples of specialist grammatical forms" (p. 395) and offers tools for analyzing discourse in ways that overcome these limitations. She analyzes the ways mathematics content is constructed by looking at what she calls the *nuclear configurations* (similar to Lemke's *thematic patterns*) in mathematics statements and oral discourse. By comparing the different ways teachers and students construct these nuclear configurations, she shows how the mathematics concepts are presented to and taken up by students. In addition, she analyzes the *reference chains* in the discourse to examine how concepts are introduced and developed and the complexity of the tracking that is needed to follow the argument being made. This entails looking at how mathematical participants in the discourse are split and recombined as a solution is developed and how the logical relationships in the text are constructed. She points out that these logical relationships can be implicit or explicit and is interested in the ambiguity that is constructed in the movement between the oral language, the mathematics symbolism, and the visual display.

O'Halloran (1999) uses grammatical analysis to compare the way mathematical meaning is made in different classroom contexts. She shows the complexity of mathematical pedagogic discourse as it moves between oral and visual modes of presentation, drawing on symbols, diagrams, and language with dense texture. Long implicational chains of reasoning based on implicit mathematics results are often used to construct mathematical knowledge with register-specific technical terms in very authoritative and sometimes alienating formations (O'Halloran, 1999).

Morgan (2004) illustrates how linguistic analysis can reveal differences in how mathematics is constructed for students at different levels by asking, *"What is the nature of mathematics/mathematical objects/mathematical activity?"* and *"Where do power and authority lie?"* (p. 6). This research helps us understand how texts used for pedagogical purposes might better recognize the nature of mathematical knowledge and negotiate its authoritative nature. She is interested in how mathematics concepts are presented to students in the texts they read, pointing out that mathematics is often presented as a set of defined concepts that can be used to generate predictable results.

This can lead to the kind of procedural pedagogy that presents mathematics as a process of learning word definitions and applying them.

Morgan is responding to mathematics standards in the UK that seem to focus only on vocabulary as a language issue and she shows how the language challenges students face go beyond word meaning. She analyzes definitions in the writing of mathematicians in academic journals and compares them with the way definitions are presented in textbooks at different levels. She shows that only textbooks present definitions as static; that in actual mathematics, definitions are presented as dynamic and evolving, open to decision-making by the mathematician. She illustrates this by comparing the way *agency* and other linguistic meanings are construed in the two kinds of texts. For example, she shows that where the mathematics research papers tend to present the authors as developing a definition (e.g., *we give a ... definition*), the student texts, especially lower level texts, present definitions without ascribing agency (e.g., *X is called...*). She finds that higher-level textbooks are more like the journal articles in allowing uncertainty and evolving meaning with a logical argument rather than absolute facts to be accepted. She suggests that ambiguity and multiplicity of meanings in classroom discussion are an important step in students' developing understanding and raises questions about the focus on vocabulary and defining terms that is common in mathematics standards and practice. She suggests that "clear explanations" may not always be the most important focus and that the more advanced texts provide better models for students of the creativity and purposefulness of mathematical practice. She argues that the model that standards set up for teachers is both too restrictive in terms of its view of language and unreflective of actual classroom practice, where teachers have to move between the technical and the more everyday language and concepts of the students.

The clause-based grammatical analysis of Veel and the discourse-based analysis of Morgan, along with the research of O'Halloran described above, are examples of how linguistic tools are available to study the "content" of the mathematics language that is constructed in classrooms at different developmental levels and in different contexts and to explore the value of different ways of presenting mathematics meaning and talking with students about those meanings. Following O'Halloran's (2005) grammatical descriptions of the mathematics register, studies could compare the depth of embedding of the key configurations students are being asked to work with, look at how the oral language that presents the same concepts is structured, and analyze the role of other semiotic systems such as the visual display and the mathematics symbolism to explore their role in students' abilities to understand and respond to the problems. The logical relationships that are built up in texts of different types, and whether they are implicit or explicit, are also of interest.

Interacting with and Positioning Students:
The Pedagogical Register

Another potentially fruitful line of research that can be pursued with linguistic tools is examination of the way the pedagogical register interacts with the content register in mathematics classrooms. This focus enables us to look at how students are positioned as they learn mathematics and the way they see themselves developing in the context of mathematics teaching. Research on the pedagogical register also can illuminate issues related to negotiating the dilemmas explored in this review in terms of how teachers can interact with students in ways that facilitate and support mathematics learning.

O'Halloran (2004) shows how linguistic tools can shed light on interpersonal relationships in the classroom and on how students are positioned as learners. Comparing differences in patterns of meaning in classrooms with students of different social classes and genders, she shows how the nature of the mathematical knowledge, the positioning of the students, and the opportunities to learn vary in relation to social class and gender variables, offering students different opportunities to engage with mathematics. She analyzes three year-10 mathematics lessons in Perth, Australia. Lesson One is taught to a class of male students at an elite school, Lesson Two is taught to female students at an elite school, and Lesson Three is taught to working class students at a public school. O'Halloran finds differences in classroom practices and the way the spoken language is used that relate to social class and gender and argues that these differences have implications for the content of the mathematics knowledge that is made available to students. She maps out the sequence of activities in each lesson, distinguishing between activities related to the mathematics content and activities that disrupt the focus on content, based on Lemke's (1990) notion of activity types. She provides visual representations of the three lessons that illustrate the "clear and progressive structure" of Lesson One in contrast with the disjointed movement between activities in Lessons Two and Three. More technical terms are used with the male elite students, and only in Lesson One is the language of the blackboard context-independent, so that it constructs a clear explanation that will make sense when copied into notebooks by the students. The lesson progresses more logically and with more internal coherence than the lessons in the other contexts. While Lesson One is focused on the mathematics content, in Lessons Two and Three the focus on content is disrupted as the teacher engages in disciplinary activities or asides. O'Halloran is able to show that less mathematics knowledge is therefore made available to students. Lesson Three uses the fewest technical terms, often moving away from the mathematics content because of disruptions due to students' behavior and the teacher's response.

O'Halloran also found that working-class and female students were "more consistently oriented towards interpersonal meaning at the expense of the mathematical content of the lesson" (2004, p. 192). The teacher and the male students in the private elite school interacted with each other directly, without the covert discipline strategies (*can I just have your attention please?*) or use of sarcasm as a control mechanism (*I just like things to be clear*) that she found in the female and working class lessons. In the male elite school, the mathematics content was foregrounded, and the interpersonal interaction was stable. In the female and working class lessons, on the other hand, the mathematics content was often backgrounded for disciplinary reasons, and the interaction of teacher with students did not show the same level of mutual respect. O'Halloran's fine-grained linguistic analysis using SFL provides evidence to support these findings. She also shows that the findings are reflected in the performance of students from these schools on national exams, with the male students scoring higher than the females who in turn score higher than the working class students. O'Halloran raises questions about how the pedagogical register of the mathematics classroom positions students in different ways, providing differential access to participation.

Morgan (2006) also considers aspects of the pedagogical register by interpreting how the meanings made in different language choices position students differently. She suggests key questions that can be asked about how mathematical knowledge is constructed and how learners are engaged with such knowledge and identifies the language features and grammatical analysis strategies that can be used to answer the questions. Seeing the language as evidence of the knowledge that is at stake as participants engage with mathematics, her goal is to interpret "the functions that these features fulfill for the participants in the mathematical practices... to gain understanding of the practices themselves" (p. 226). She shows how the language constructs students as actively *doing* in some cases and the mathematics as just *given* in others (e.g., the difference between *if you do this, X increases* and *here is the formula*). She argues that these differences in wording "construct different images of the objects of mathematics and the nature of mathematical activity. At the same time they claim different types of authority and construct different 'ideal' positions for their readers" (p. 236). She also points out that looking at the actual language used in mathematics teaching and learning can help researchers recognize teachers' and students' beliefs about the nature of mathematics without relying on self-reports or responses to explicit or implicit questioning outside the context of actually doing mathematics.

Morgan suggests, for example, that when textbooks use procedural discourse as their organizational strategy, they "may make students more likely to perceive mathematics as consisting of a set of procedures, and hence, perhaps, to find it more difficult to engage with relational or logical aspects of the subject" (2006, p. 228). Texts that obscure human agency "may contribute to difficulties for some students in seeing themselves as potential mathematicians" (p. 228). She illustrates this by comparing the texts written by two students in response to a problem, looking at how they represent mathematical objects, the processes they are involved in, and who is acting in the processes. By observing the patterns constructed by the students, she is able to show that one student draws "primarily on a discourse of investigation, oriented to value exploration of interesting mathematics," while the other "draws strongly on an assessment discourse, displaying the 'answers' valued within that discourse" (p. 236). By looking at the text as a whole, she can track changes in it as it progresses and is negotiated and show how meaning evolves. Morgan calls for more such analysis to show how texts construct different images of mathematics and how those are read by students, and how students adopt and reproduce the images of mathematics constructed by their teachers. A linguistic analysis can also show differences in teaching styles and student participation, the effect of resistance by students on teachers' evaluation of them, and how the social positioning of students in the classroom constructs their identities as effective or less effective mathematics students. Morgan suggests that knowledge about how various responses are evaluated is especially important in the current high stakes testing environment.

Research can also show how classroom practices in mathematics can lead to student alienation. In interviews with 12 first-year university students, Solomon (2006) found that they did not see themselves as potential members of the mathematics discourse community, even when they were majoring in mathematics and interested in the subject. She blames this on a mode of teaching that focuses on procedures and rote learning rather than negotiation of meaning and development of understanding through participation. She suggests that "learners...often are excluded from the negotiation of meaning...developing instead an identity of non-participation and marginalisation. Their lack of ownership generates and is generated by compliance with authority and an emphasis on following pre-set procedures" (pp. 376–377). She points out that while professional mathematics involves procedures such as "*intuition, trial, error, speculation, conjecture, proof,*" undergraduate teaching of mathematics stresses "*definition, theorem, proof,*" and as a consequence, students see mathematics learning as rote learning and see proof as instrumental and performance-related. As she summarizes her findings, "[a] general theme of certainty in mathematics emerged, coupled with an emphasis on the necessity of learning rules, reproducing solutions and working at speed to get correct answers" (Solomon, 2006, p. 382). The students assumed "certainty, irrelevance, rule-boundedness and lack of creativity potential in pure mathematics" (p. 383) and were "largely unaware of the existence of a mathematics community of practice which might have negotiable rules of communication and validation" (p. 384). She suggests that the way mathematics is taught means that students are unlikely to be able to develop an identity as a mathematician, and that students need to "make the transition from a performance-oriented and individualistic view of mathematics... to a view... which emphasises construction, communication and community" (p. 391).

These studies suggest promising directions for research on the pedagogical register and student positioning in mathematics classrooms that may help untangle issues related to the different ways students encounter mathematics in today's multicultural classrooms and help teachers address the tensions they confront as they negotiate language and mathematics learning. By identifying features of the pedagogical register that facilitate the development of students' understanding, the role of language in talking about mathematical meaning can be better understood, and ways for teachers and students to talk about mathematics at different developmental levels and in different topics can be elucidated.

Intertextuality and the Multi-Semiotic Nature of Mathematics

Two other key issues also need further study in mathematics teaching and learning. One is movement between and among texts of different types, spoken and written, in classroom instruction, as students listen to the teacher, work with textbooks, and interact with the blackboard and other visual representations. Related to this is the need for more research on the role of the multi-semiotic nature of the mathematics content register in learning and teaching. The movement between and among texts of different types has been addressed in some research on intertextuality in mathematics. Studies have compared the language used by teachers and students or by students and textbooks in talking about the same concepts (Chapman, 1995; Veel, 1999), and these studies find that the students do not readily take up the language in the same ways that the teacher or the textbook use it, with implications for the learning that is taking place. Less work has been done on the multi-semiotic nature of mathematical discourse.

Huang, Normandia, and Greer (2005), for example, following a line of research based on SFL and Mohan (1986), analyze the knowledge structures that appear in teacher and in student discourse in a mathematics class where the teacher expected technical language from the students and encouraged them to talk about mathematics. They report that students, working in groups

could easily describe an equation or a graph, sequentially tell about procedures they have followed to solve a function, and suggest a method or solution. However, whenever students were pushed to reference relevant concepts
or principles, explain a method used, or justify a decision made for either a
method or solution, they frequently seemed to hesitate or to appear less capable. (Huang et al., 2005, p. 44)

Huang and colleagues found that knowledge structures such as *classification,
principles,* and *evaluation* were only used by the teacher. Students used only
description, sequence, and *choice* knowledge structures even when pushed by
the teacher. The authors suggest that students need explicit instruction in
articulating principles to move them beyond the practical aspects of mathematics knowledge in their discussion. They recommend that students be
asked to "talk their way into habits of expressing higher-level knowledge
structures" (pp. 44–45) and that teachers integrate thinking and talking at
all levels. Borasi and Siegel (2000) also make connections between reading,
talk, and drawing in geometry and statistics and offer strategies for making
links across modalities.

Chapman's work (1995, 2003a, 2003b), based on an ethnographic study
of a year nine mathematics class in Australia, offers a rich model of how
intertextuality can be studied. She is interested in language use by students
and teachers as well as the role of texts, diagrams, and other visuals interacting with the mathematics symbolism. She shows how knowledge is built
up as students interact with spoken and written "texts" over the course of
a unit on functions. (The discussion of thematic formations, above, draws
from this work.) With a social semiotic perspective influenced by Lemke
and SFL, she analyzes the semantic and thematic patterns and genre structures of the mathematics class, showing how the different texts in a unit of
study offer opportunities for students to learn intertextually, and that in
fact such learning is necessary, as no one element captures the entirety of
the concepts that the unit develops. She shows that the thematic formations
that construct the knowledge evolve through different patterns of expression, each pattern adding to the common pattern of expression that the
teacher and students are building up.

Chapman (1995) shows how understanding is built up over the course
of the whole unit as she analyzes the ways the teacher and students interact
in whole group sequences, looks at the thematic patterns in teacher talk,
analyzes the students' interaction in small groups, and looks at how the
textbook represents the same thematic patterns. She shows that the teacher
focuses students on using appropriate language to construct concepts and
that she contextualizes the textbook language through spoken language
to help students understand the relationships in the textbook definitions.
Chapman also shows that the definitions in the textbook assume students
already understand the relationships among what is defined. She illustrates

that when the students work out a problem together in groups, some construct the concepts in ways that are different from the thematic patterns of the teacher and the textbook, sometimes leading to miscommunication as students use different thematic formations to talk about the mathematics content. Chapman suggests that the intertextual links can be highlighted for students if the teacher makes explicit reference to the key thematic patterns and relations, using ones that are familiar in making connections with other texts and other contexts, so that the thematic items are naturalized into the language of the classroom.

This work shows how mathematics concepts are developed through engagement with and talk about texts of different types, including homework, blackboard diagrams, the textbook, teacher–student interaction, and students' interaction with each other. It illustrates the role the teacher's language plays, showing how the teacher's language only makes sense in relation to the other texts the class engages with. Chapman also shows that mathematics texts require a different kind of reading than texts in other subjects, as they depend on the completion of exercises and classroom discussion for students to understand the definitions and other meanings that present the thematic patterns being developed. Chapman urges that teachers explicitly reference the thematic patterns under construction as they transform nonmathematical expressions into mathematical ones.

In other work, Chapman shows that developing the mathematical discourse is a developmental process that evolves as students gain experience with new mathematical knowledge. Chapman (2003b) stresses that meaning is always produced in context and cannot be separated from social action and that "ways of speaking that are appropriate to a subject-area are developed as part of the social practices of classroom interaction" (p. 133). Her notion of social context draws on Lemke in pointing out its three dimensions: the content context, or what is being talked about; the sequential context, related to what has come before and what will come after; and the context of what has been selected to highlight in each instance, in contrast to what other elements might have been selected and highlighted. She suggests that these three angles on mathematics content help us see where teachers are using everyday language in metaphoric ways to represent mathematics. When she analyzes the relevance of the metaphoric content to the problem at hand, she finds that the more metaphoric ways of representing mathematics do not always represent the mathematics content accurately, as it is in the mathematics language itself that the mathematical meaning is made.

We need more research that compares texts of different types, reveals problems with different wording, and analyzes representations of mathematics at different levels and in different topics. Most needed is research that shows effective ways of negotiating the multi-semiotic nature of mathematics. We need to understand how teachers "translate" across the different semiotics, identify where ambiguity occurs and its value, and develop more explicit descriptions of how language works across modalities. We need to know more about the language patterns and structures used in talking about visual elements in mathematics classrooms and the natural language used to interpret the mathematics symbols.

Summary

A functional linguistics perspective offers new ways of analyzing language to explore both how mathematics knowledge is constructed and how teachers initiate students into that knowledge in different contexts and at different levels. Studies that consider the actual language used in mathematics classrooms from a multi-semiotic perspective that recognizes the roles of language, mathematics symbolism, and visual display in constructing mathematical meaning can illuminate the issues this review has raised about movement between everyday and technical ways of making mathematical meaning and how best to engage students in diverse contexts. Analysis of the texts used for instruction and the ways teachers and students use and respond to those texts can identify challenges in the language and suggest ways to enable more effective presentation of content. Through comprehensive investigation of how technical meanings are made, the language structures that enable meaning, forms of interaction in the classroom, description of spoken and written genres, and the ways multi-semiotic meaning-making works in mathematics teaching and learning, the language resources that are most relevant in constructing different mathematics topics at different levels can be identified and made explicit.

THE ROLE OF TEACHER EDUCATION

Mathematics teachers need more than knowledge about mathematics. Ball and Bass (2003) argue that knowledge about how to link content and pedagogy, what they call *mathematical knowledge for teaching*, is key to preparing effective teachers. Teachers need to know how to respond to children who offer alternative solutions, how to make mathematical ideas available to students, how to attend to, interpret, and handle students' oral and written productions, how to give and evaluate mathematical explanations and justifications, and how to establish and manage the discourse and collectivity of the class for mathematics learning (Ball & Bass, 2003).

Because teachers need to do interpretive work in developing and responding to students' learning in mathematics classrooms, Ball, Hill, and Bass (2005) see the teaching of mathematics as a form of mathematical work that requires "a special sort of sensitivity to the need for precision in mathematics . . . that language and ideas be meticulously specified so that mathematical problem solving is not unnecessarily impeded by ambiguities of meaning and interpretation" (p. 8). They report that they "have seen students struggle over language, where terms were incompletely or inconsistently defined, and we have seen discussions which run aground because mathematical reasoning is limited by a lack of established knowledge foundational to the point at hand" (p. 12) and suggest that "an emergent theme in our research is the centrality of mathematical language . . . teachers must constantly make judgments about how to define terms and whether to permit informal language . . . When might imprecise or ambiguous language be pedagogically preferable and when might it threaten the development of correct understanding?" (p. 21).

This suggests that teaching mathematics is also a form of *linguistic work*. New registers evolve for students along with the development of their mathematics knowledge. As they do mathematics, they come to understand it in new ways. The challenge is that students have to use the mathematics register while learning the register. Teachers need to understand mathematics concepts well enough to be able to talk about them in both less and more technical ways, unpacking compressed conceptions or formal definitions of the technical language into the more informal, everyday language for student learning, but then compressing and repackaging it again in technical ways in order for the learning to build more complex concepts and move into new, more abstract domains. It is the movement back and forth that allows students to make connections across mathematical topics and for teachers to build links and establish coherence as learning develops.

Professional development can be a powerful venue for examining and disseminating new understanding of the relationship between language and mathematics so that we can better prepare teachers to deal with the tensions around language and content, raising teachers' awareness about language, giving them ways of explicitly talking about language features and modeling ways of responding to students' alternative explanations and questions. In the context of teacher development, a focus on language can help address the difference between questions like *do you know mathematics?* and *can you talk about mathematics?*, providing teachers with ways of engaging students in talk about the knowledge they are developing that maintain the technicality and structure necessary for the concepts to be effectively engaged with.

The linguistic framework suggests ways of responding to the dilemmas identified by researchers by enabling talk about meaning: both talk about *what it means* and *how it means*, providing the potential for recognizing how meaning is constructed in mathematics and in mathematics classrooms. The SFL metalanguage (a language about language) can help negotiate the tensions, making the forms through which mathematics knowledge is presented explicit and providing a valuable tool that can be developed and studied in various contexts of use.

We know from research on school facilities and teacher quality that privileged students have more educational resources and achieve better outcomes than less privileged students. Understanding the linguistic challenges of mathematics may offer new ways of improving instruction for all students despite these structural obstacles. The studies reviewed here illustrate the variety and richness of the insights available through functional linguistics tools, with methodological possibilities for research ranging from ethnographic studies to grammatical analyses and discourse analysis. More such research can enhance our understanding of the role language plays in mathematics education to help better understand what constitutes the mathematical knowledge for teaching that is most important.

RECOMMENDATIONS OF THIS REVIEW

Research on classroom talk that shows effective ways of building on students' everyday language and developing their academic language can help us understand which areas of language are most readily addressed by teachers explicitly and what the outcomes are for students whose teachers are able to do this. A key issue is the complexity of using the natural language, mathematics symbolism, and visual display that make up the multi-semiotic mathematical discourse. How explicit should a teacher be, and how much everyday language and technical language should be used? Related to this is the issue of how to move students into a field that objectifies concepts and presents them in dense language that can be difficult to engage with. Should the students be given the formal definitions that present the knowledge mathematicians have developed, or should students be taken through a process that moves from less precise to more precise construction of knowledge? And how should teachers address the alienating quality of mathematical discourse and its authoritative constructions?

We need studies that illuminate the features of effective pedagogical and mathematics content registers in order to develop the metalanguage that is useful for teachers and students at different levels and on different mathematics topics. Having a metalanguage that effectively supports students' development of mathematics concepts in different learning contexts can help teachers move beyond the everyday/technical dichotomy and start with the language students bring to school, but then move them into the ways of using language that effectively construct the mathematical knowledge they need to develop.

We also need to analyze more fully the multi-semiotic nature of the construction of mathematics knowledge and investigate how effective teachers shift among the different modes and semiotic systems as they negotiate movement between oral language, mathematics symbolism, and visual display. This needs to be studied in different contexts and at different levels with various mathematics topics. Research that describes teachers' language and compares teachers' and students' ways of talking about mathematics can help us better understand the gap that has been seen between student use of mathematical language and teacher/textbook use of mathematical language and where the gap can be bridged. Studies of different kinds of texts, different ways of interacting in classrooms, and pathways that students' mathematical development takes can also be undertaken using linguistic tools that enable analysis of meaning-making in mathematics.

We need to look at different kinds of classrooms, situated in different social formations, to compare how students are positioned through the mathematics discourse and how teacher–student interaction varies across settings and the implications of this for teaching and learning. We need to understand differences in linguistic behavior among different language users and social contexts in the mathematics classroom and to compare teacher use of language in different contexts. Diverse contexts of research are necessary, as we cannot expect that one way of interacting with students will be effective with all. Research needs to include second language learners and speakers of non-standard dialects in a range of social contexts as well. It would be especially valuable to focus on settings where teachers have been successful to see how successful teachers use language in mathematics teaching in different learning contexts.

More research is needed that takes a developmental approach and shows how students develop knowledge over time. We need rich studies of how language and ways of talking about mathematics evolve over a unit of study, focusing on more than brief interactional episodes and fragments of dialogue. We need research that describes developmental paths into K–12 mathematics at different grades and on different topics for students in different contexts and with different backgrounds. Such research can also investigate how a metalanguage that enables connections to be made between language and mathematics meanings can support students' learning.

To make changes in the ways teachers use and think about language in mathematics classrooms, we need research that investigates the kind of understanding about language that it is possible for teachers to take up at different points in their development. Rich studies of mathematics teacher preparation where knowledge about mathematical discourse is made available to teachers and implemented in classrooms at different levels and in different contexts can provide information about the ways of using language that are most effective in helping students learn mathematics. Such studies should follow teachers' development and measure their performance. This is a crucial area for research, and the new linguistic approaches have much to offer.

Semiotic approaches to mathematics education that recognize that knowledge is constructed in language and that mathematics is especially challenging in its multimodal construal have much to offer researchers. Social semiotic approaches that recognize the different ways students from different cultures, language backgrounds, and social contexts use language offer especially rich opportunities to develop new approaches to mathematics education that recognize and value different ways of learning and using language. The dilemmas and challenges are clear, and the methodological tools are available. A robust program of research that approaches these issues in systematic and deep ways can make major contributions to the field of mathematics education through the study of language use.

REFERENCES

Abedi, J., & Lord, C. (2001). The language factor in mathematics tests. *Applied Measurement in Education, 14*(3), 219–234.

Adler, J. (1997). A participatory-inquiry approach and the mediation of mathematical knowledge in a multilingual classroom. *Educational Studies in Mathematics, 33*, 235–258.

Adler, J. (1998). A language of teaching dilemmas: Unlocking the complex multilingual secondary mathematics classroom. *For the Learning of Mathematics, 18*(1), 24–33.

Adler, J. (1999). The dilemma of transparency: Seeing and seeing through talk in the mathematics classroom. *Journal of Research in Mathematics Education, 30*(1), 47–64.

Anderson, M., Sáenz-Ludlow, A., Zellweger, S., & Cifarelli, V. V. (Eds.). (2003). *Educational perspectives on mathematics as semiosis: From thinking to interpreting to knowing.* Ottawa: Legas.

Ball, D. L., & Bass, H. (2003). Toward a practice-based theory of mathematical knowledge for teaching. In B. Davis & E. Simmt (Eds.), *Proceedings of the 2002 annual meeting of the Canadian Mathematics Education Study Group* (pp. 3–14). Edmonton, AB: CMESG/GCEDM.

Ball, D. L., Hill, H. C., & Bass, H. (2005). Knowing mathematics for teaching. *American Educator, Fall,* 14–46.

Barwell, R. (2003). Patterns of attention in the interaction of a primary school mathematics student with English as an additional language. *Educational Studies in Mathematics, 53,* 35–59.

Barwell, R. (2005a). Integrating language and content: Issues from the mathematics classroom. *Linguistics and Education, 16,* 205–218.

Barwell, R. (2005b). Working on arithmetic word problems when English is an additional language. *British Educational Research Journal, 31*(3), 329–348.

Barwell, R., Leung, C., Morgan, C., & Street, B. (2005). Applied linguistics and mathematics education: More than words and numbers. *Language and Education, 19*(2), 141–146.

Bernstein, B. (1996). *Pedagogy, symbolic control and identity: Theory, research, critique.* London: Routledge & Kegan Paul.

Borasi, R., & Siegel, M. (2000). *Reading counts: Expanding the role of reading in mathematics classrooms.* New York: Teachers College Press.

Brown, C. L. (2005). Equity of literacy-based math performance assessments for English language learners. *Bilingual Research Journal, 29*(2), 337–363.

Chapman, A. (1995). Intertextuality in school mathematics: The case of functions. *Linguistics and Education, 7*(3), 243–362.

Chapman, A. P. (2003a). *Language practices in school mathematics: A social semiotic approach.* Lewiston, NY: Edwin Mellen Press.

Chapman, A. P. (2003b). A social semiotic of language and learning in school mathematics. In M. Anderson, A. Sáenz-Ludlow, S. Zellweger, & V. V. Cifarelli (Eds.), *Educational perspectives on mathematics as semiosis: From thinking to interpreting to knowing* (pp. 129–148). Brooklyn, NY and Ottawa, Ontario: Legas.

Christie, F. (1991). First- and second-order registers in education. In E. Ventola (Ed.), *Functional and systemic linguistics* (pp. 235–256). Berlin: Mouton de Gruyter.

Christie, F. (2002). *Classroom discourse analysis: A functional perspective.* London: Continuum.

Cobb, P., & Yackel, E. (1998). A constructivist perspective on the culture of the mathematics classroom. In F. Seeger, J. Voigt, & U. Waschescio (Eds.), *The culture of the mathematics classroom* (pp. 158–190). Cambridge: Cambridge University Press.

Cobb, P., Yackel, E., & McClain, K. (Eds.). (2000). *Symbolizing and communicating in mathematics classrooms: Perspectives on discourse, tools, and instructional design.* Mahwah, NJ: Lawrence Erlbaum Associates.

Forman, E. A., McCormick, D. E., & Donato, R. (1997). Learning what counts as a mathematical explanation. *Linguistics and Education, 9*(4), 313–340.

Gerofsky, S. (1996). A linguistic and narrative view of word problems in mathematics education. *For the Learning of Mathematics, 16*(2), 36–45.

Gorgorió, N., & Planas, N. (2001). Teaching mathematics in multilingual classrooms. *Educational Studies in Mathematics, 47*, 7–33.

Halliday, M. A. K. (1978). *Language as social semiotic.* London: Edward Arnold.

Halliday, M. A. K., & Matthiessen, C. M. I. M. (2004). *An Introduction to Functional Grammar* (3rd ed.). London: Arnold.

Herbst, P. (2003). Using novel tasks in teaching mathematics: Three tensions affecting the work of the teacher. *American Educational Research Journal, 40*(1), 197–238.

Huang, J., B. Normandia, & S. Greer. (2005). Communicating mathematically: Comparison of knowledge structures in teacher and student discourse in a secondary math classroom. *Communication Education, 54*(1), 34–51.

Khisty, L. L., & Viego, G. (1999). Challenging conventional wisdom: A case study. In L. Ortiz-Franco, N. Hernandez, & Y. De La Cruz (Eds.), *Changing the Faces of Mathematics* (pp. 71–80). Reston, VA: National Council of Teachers of Mathematics.

Laborde, C., Conroy, J., De Corte, E., Lee, L., & Pimm, D. (1990). Language and mathematics. In P. Nesher & J. Kilpatrick (Eds.), *Mathematics and cognition: A research synthesis by the International Group for the Psychology of Mathematics Education* (pp. 53–69). Cambridge, UK: Cambridge University Press.

Lager, C. A. (2006). Types of mathematics-language reading interactions that unnecessarily hinder algebra learning and assessment. *Reading Psychology, 27,* 165–204.

Lampert, M. (1985). How do teachers manage to teach? Perspectives on problems in practice. *Harvard Educational Review, 55*(2), 178–194.

Lemke, J. (1990). *Talking science: Language, learning, and values.* Norwood, NJ: Ablex.

Lemke, J. L. (2003). Mathematics in the middle: Measure, picture, gesture, sign, and word. In M. Anderson, A. Sáenz-Ludlow, S. Zellweger, & V. V. Cifarelli (Eds.), *Educational perspectives on mathematics as semiosis: From thinking to interpreting to knowing* (pp. 215–234). Brooklyn, NY and Ottawa, Ontario: Legas.

MacGregor, M. (2002). Using words to explain mathematical ideas. *Australian Journal of Language and Literacy, 25*(1), 78–88.

Marks, G., & Mousley, J. (1990). Mathematics education and genre: Dare we make the process writing mistake again? *Language and Education, 4*(2), 117–135.

McNamara, O. (2003). Locating Saussure in contemporary mathematics education discourse. In M. Anderson, A. Sáenz-Ludlow, S. Zellweger, & V. V. Cifarelli (Eds.), *Educational perspectives on mathematics as semiosis: From thinking to interpreting to knowing* (pp. 17–34). Ottawa: Legas.

Mohan, B. A. (1986). *Language and content.* Reading, MA: Addison-Wesley.

Morgan, C. (2004). Word, definitions and concepts in discourses of mathematics, teaching and learning. *Language and Education, 18,* 1–15.

Morgan, C. (2006). What does social semiotics have to offer mathematics education research? *Educational Studies in Mathematics, 61,* 219–245.

Moschkovich, J. N. (1999). Supporting the participation of English language learners in mathematical discussions. *For the Learning of Mathematics, 19*(1), 11–19.

Moschkovich, J. N. (2000). Learning mathematics in two languages: Moving from obstacles to resources. In W. Secada (Ed.), *Changing the faces of mathematics (Vol. 1): Perspectives on multiculturalism and gender equity* (pp. 85–93). Reston, VA: National Council of Teachers of Mathematics (NCTM).

Moschkovich, J. N. (2002). A situated and sociocultural perspective on bilingual mathematics learners. *Mathematical Thinking and Learning, 4*(2 & 3), 189–212.

Nunes, T., Schliemann, A. D., & Carraher, D. W. (1993). *Street mathematics and school mathematics.* Cambridge: Cambridge University Press.

O'Halloran, K. L. (1999). Towards a systemic functional analysis of multisemiotic mathematics texts. *Semiotica, 124*(1/2), 1–29.

O'Halloran, K. L. (2000). Classroom discourse in mathematics: A multisemiotic analysis. *Linguistics and Education, 10*(3), 359–388.

O'Halloran, K. L. (2004). Discourses in secondary school mathematics classrooms according to social class and gender. In J. A. Foley (Ed.), *Language, education and discourse: Functional approaches* (pp. 191–225). London: Continuum.

O'Halloran, K. L. (2005). *Mathematical discourse: Language, symbolism and visual images*. London: Continuum.

Pimm, D. (1987). *Speaking mathematically: Communication in mathematics classrooms*. London: Routledge & Kegan Paul.

Presmeg, N. C. (1998). A semiotic analysis of students' own cultural mathematics. Research Forum Report. In A. Olivier & K. Newstead (Eds.), *Proceedings of the 22nd Conference of the International Group for the Psychology of Mathematics Education, 1*, 136–151.

Radford, L. (2000). Signs and meanings in students' emergent algebraic thinking: A semiotic analysis. *Educational Studies in Mathematics, 42*, 237–268.

Sáenz-Ludlow, A. (2003). Classroom discourse in mathematics as an evolving interpreting game. In M. Anderson, A. Sáenz-Ludlow, S. Zellweger, & V. V. Cifarelli (Eds.), *Educational perspectives on mathematics as semiosis: From thinking to interpreting to knowing* (pp. 253–281). Ottawa: Legas.

Schleppegrell, M. J. (2007). The linguistic challenges of mathematics teaching and learning: A research review. *Reading & Writing Quarterly, 23*, 139–159.

Selin, H., & D'Ambrosio, U. (Eds.). (2000). *Mathematics across cultures: The history of non-western mathematics*. Dordrecht, Boston: Kluwer Academic.

Setati, M. (2005). Teaching mathematics in a primary multilingual classroom. *Journal for Research in Mathematics Education, 36*(5), 447–466.

Setati, M., & Adler, J. (2001). Between languages and discourses: Language practices in primary multilingual mathematics classrooms in South Africa. *Educational Studies in Mathematics, 43*, 243–269.

Sfard, A. (2001). There is more to discourse than meets the ears: Looking at thinking as communicating to learn more about mathematical learning. *Educational Studies in Mathematics, 46*, 13–57.

Sfard, A., & Lavie, I. (2005). Why cannot children see as the same what grown-ups cannot see as different?–Early numerical thinking revisited. *Cognition and Instruction, 23*(2), 237–309.

Sfard, A., Nesher, P., Streefland, L., Cobb, P., & Mason, J. (1998). Learning mathematics through conversation: Is it as good as they say? *For the Learning of Mathematics, 18*(1), 41–51.

Silver, E. A., & Smith, M. S. (1996). Building discourse communities in mathematics classrooms: A worthwhile but challenging journey. In P. Elliott & M. Kenney (Eds.), *1996 Yearbook—Communication in mathematics, K–12 and beyond* (pp. 20–28). Reston, VA: NCTM.

Solomon, Y. (1989). *The practice of mathematics*. London: Routledge.

Solomon, Y. (2006). Deficit or difference? The role of students' epistemologies of mathematics in their interactions with proof. *Educational Studies in Mathematics, 61*, 373–393.

Solomon, Y., & O'Neill, J. (1998). Mathematics and narrative. *Language and Education, 12*(3), 210–221.

Staub, F. C., & Reusser, K. (1995). The role of presentational structures in understanding and solving mathematical word problems. In C. A. Weaver, S.

Mannes, & C. R. Fletcher (Eds.), *Discourse comprehension* (pp. 285–305). Hillsdale, NJ: Lawrence Erlbaum Associates.

Veel, R. (1999). Language, knowledge and authority in school mathematics. In F. Christie (Ed.), *Pedagogy and the shaping of consciousness: Linguistic and social processes* (pp. 185–216). London: Continuum.

Zack, V., & Graves, B. (2001). Making mathematical meaning through dialogue: "Once you think of it, the Z minus three seems pretty weird." *Educational Studies in Mathematics, 46,* 229–271.

CHAPTER 4

FUNCTION AND FORM IN RESEARCH ON LANGUAGE AND MATHEMATICS EDUCATION

Guillermo Solano-Flores

ABSTRACT

My goal in this chapter is to examine the intersection of language and mathematics in educational research from a multidisciplinary perspective. My research addresses mathematics and science testing for English language learners in large-scale testing programs (e.g., NAEP) and the testing of linguistically diverse populations in international test comparisons (e.g., PISA and TIMSS). I develop assessment models and strategies that are informed by current knowledge from the language sciences. In addition to my formal training in psychometrics—the discipline that deals with the measurement of behavior and cognition in humans and social systems—my work has given me exposure to sociolinguistics, cognitive science, second language acquisition, translation, and cultural anthropology, among other fields. I have become aware that researchers from different specialties in education vary tremendously on how they think about language and what is necessary and acceptable practice to properly address issues of language in education. While those views are not necessarily incompatible, professionals tend to use only one view

Language and Mathematics Education, pages 113–149

to inform their work. In this chapter, I submit the notion that four views of language—as a process, system, structure, and factor—shape how researchers of different orientations investigate language and its relation to mathematics. In order to attain more valid and fair mathematics testing practices for linguistically diverse populations, we need to learn to both use multiple views of language in combination and communicate more effectively across areas of specialty.

INTRODUCTION

The symbolic nature of mathematics is intriguing to many—to the extent that popular conceptions exist about the linguistic nature of mathematics. One conception is that mathematics is a language in its own right, a universal language; another conception is that mathematics is language-free. In mathematics education, David Pimm (1987) provided a balanced analysis of the linguistic nature of mathematics. As Pimm explained, while mathematics has properties that can be examined from the perspective of linguistics, mathematics is not a language that can be learned in the same way people learn foreign languages. At the same time, natural, ordinary language can be used as a communicative tool for interpreting and constructing meaning in mathematics. As in any discipline or social activity, there is a register and a set of conventions that are specific to mathematical communication.

Researchers and practitioners in the field of mathematics education face language issues and need to be aware of the multiple aspects of the complex relationship between language and mathematics learning. For example, the *Curriculum and Evaluation Standards for School Mathematics* (National Council of Teachers of Mathematics, 1989) identify "learning to communicate mathematically" as a major goal for all students. A later publication, the *Principles and Standards for School Mathematics* (National Council of Teachers of Mathematics, 2000) also recognizes the importance of organizing and consolidating mathematical thinking through communication and the importance of using the "language of mathematics to express mathematical ideas" (p. 60). The second document also includes among its goals addressing the instructional needs of students with limited proficiency in the language of instruction.

Congruence in research on the teaching and assessment of mathematics is possible only if there is clarity about the multiple ways in which "communication," "language of mathematics," "limited language proficiency," "language of instruction," and other language-related concepts are understood, studied, and measured. This congruence is especially important in times of standards and accountability. Alignment with standards documents is regarded as evidence of good practice and test validity (see Schoenfeld, 2004; Sloane & Kelly, 2003). Regardless of whether or not we agree with this trend, we cannot help but wonder whether the goals underlying standards documents can be accomplished when teaching and mathematics assessment are guided by different conceptions of language. We can also wonder whether language is addressed in the same ways by researchers who deal with language in the classroom and those who deal with language in testing. Understanding these differences is critical to both ensuring proper interpretation of results across research areas and improving mathematics teaching and assessment research and practice in relation to language. It also may be helpful in identifying promising areas for future research on language and mathematics and in generating knowledge that can inform practice and policy (see Burkhardt & Schoenfeld, 2003).

In this chapter, I address the notion that both implicit and explicit conceptions of language in mathematics education influence which variables researchers regard as relevant, what kind of data they gather, and how they interpret their findings (Schoenfeld, 2007). I offer a conceptual framework for examining research on language and mathematics education. I submit that research on language and mathematics education is guided by views of language—ways in which the phenomenon of language is conceived.

The first section of the chapter presents an overview of the conceptual framework. While they are often implicit, views of language guide the reasonings and assumptions researchers use in their investigations. Also, while these language views are not necessarily antithetical or mutually exclusive, researchers tend to adhere mostly to one view in their work. The second and third sections examine two broad types of views that I identify in educational research on language and mathematics education, functional and formal. Functional views encompass views of language as a process and as a system; formal views encompass views of language as a structure and as a factor. Some examples are provided (although not discussed in detail) with the intent to illustrate the wide variety of investigations that may share each view. These investigations may reflect opposing theoretical perspectives in mathematics education. In the last section, I present a few recommendations and discuss promising areas for research on this topic.

OVERVIEW OF LANGUAGE VIEWS IN RESEARCH ON LANGUAGE AND MATHEMATICS

Research on language and mathematics education can be characterized according to four language views: as a process, as a system, as a structure, and as a factor. None of these language views is better than the others. While they emphasize different aspects of language, they are not antithetical or mutually exclusive.

Table 4.1 compares these four views of language along seven dimensions: (1) roles attributed to language as a key actor in mathematics educa-

TABLE 4.1 Views of Language in Research on Language and Mathematics

Dimensions	Functional views		Formal views	
	As a process	As a system	As a structure	As a factor
1. Role of language	Means for understanding	Resource for knowledge construction	Agent of problem complexity	Extraneous variable
2. Themes or ideas	Development and cognition	Social interaction and communication	Organization and difficulty	Condition and control
3. Unit(s) of analysis	Individual learner, classroom	Classroom, community	Problem type	Group
4. Language modes	Speaking and writing	Classroom conversation	Reading (printed text)	Reading and writing
5. Key concepts	Meaning, register	Language, dialect, discourse, bilingualism, multilingualism, language choice, language contact	Grammar constituents, semantic structure	Language proficiency, testing conditions, test bias
6. Broad areas of research	Language influences in the development of mathematical knowledge	Linguistic diversity in mathematics education	Influence of the linguistic features of mathematics problems on student performance and problem solving strategies	Effect of language differences on the accuracy of measures of mathematics achievement
7. Theories and disciplines	Sociocultural theory, constructivism, cultural-historical activity theory, discourse theory, cognitive psychology, cognitive anthropology, cultural anthropology, sociolinguistics		Psychometrics, item response theory, cognitive psychology, structural linguistics	

tion, (2) themes or ideas underlying the researchers' actions, (3) units of analysis, (4) language modes most frequently used as sources of data, (5) key concepts, (6) broad areas of research, and (7) theories and disciplines most commonly used. The table is intended to show patterns rather than clear-cut distinctions between language views. A given investigation may address language in ways that are typical to one view for some dimensions and to another view for other dimensions. However, researchers tend to adhere in their investigations to one view for most of the dimensions, if not all.

Views of language as a process and as a system can be referred to as *functional* because they emphasize the dynamic aspect of language in mathematical communication and the development of mathematical knowledge. Views of language as a structure and as a factor can be called *formal* because they emphasize linguistic features of mathematical problems or types of linguistic groups.

FUNCTIONAL VIEWS

Language as a Process

Views of language as a process are observed in research that examines the role of language in the development of mathematical knowledge. Language can be examined as a reflection of mathematical understanding based on the analysis of verbal reports of students as they solve mathematical problems (e.g., Siegler & Jenkins, 1989), for example, with the purpose of determining how the ways in which children use number words in different contexts reveal aspects of their understanding of the notion of cardinality (see Fuson, 1991). Language as a social process (see Vygotsky, 1936/1986) and culture as a phenomenon that shapes mind (Vygotsky, 1978; Wertsch, 1985) are seen as key in the development of mathematical thinking (see Brenner, 1994; Lampert, 1990). Communication is examined as both a facilitator of learning—communicating to learn mathematics— and a learning goal—learning to communicate mathematically—(Lampert & Cobb, 2003) in teacher–student and student mathematical conversations (e.g., Brenner, 1998a; Khisty, 1995; Webb, 1991). Also, cultural differences in mathematical reasoning are examined to understand the role of culture and cultural identity in the learning and mental representation of mathematics (Garcia, 1993; Ladson-Billings, 1995; Nasir, 2002; Stigler & Baranes, 1988–1989).

Here I discuss two aspects of language as a process: preserving meaning—how language shapes meaning when it is transferred across mathematics and natural language—and negotiating meaning—constructing mathematical knowledge through social interaction.

Preserving Meaning

Brown and Yule (1983) propose the existence of two functions of language, transactional (transferring information) and interactional (establishing and maintaining social relationships). The transactional function of language is of special interest in mathematics education because it is concerned with the correspondence between natural language and mathematics.

Some scientific concepts may be difficult to learn because the words used to refer to them have different meanings in everyday life (e.g., Meyerson, Ford, Jones, & Ward, 1991). In their attempts to make sense of mathematics word problems, students may substitute words (e.g., *for every—for each; through—in*) when they rephrase them (Mitchell, 2001). Due to the overlap of meaning in the discipline and in the natural language, students need to be familiar with the meaning of words at a level of understanding that goes beyond the knowledge passed on by definitions (Nagy, 1988).

The notion of register is critical to examining how the characteristics of natural language shape students' interpretations of mathematical representations. *Register* is a term that refers to the fact that written or spoken language varies across situations and activities (Halliday, 1978) and, more specifically, to the ways in which certain things and concepts (e.g., *integer, subtract*) are referred to by a community (e.g., the community of mathematicians or the community of mathematics educators) as a result of a social process that involves specialization in a content area, certain contexts, and certain specific activities.

As with science, writing in mathematics is distinguished from ordinary writing by virtue of a high frequency of features such as nominalization, impersonal style, passive voice, lexical density, and the use of interlocking definitions, among many others (Halliday, 1993; Morgan, 1998, 2005). Altogether, these features make it difficult for students to understand technical writing. For example, *glass crack growth rate* may be more difficult to understand than *how quickly cracks in glass grow* (Halliday, 1993, p. 79), which conveys the same meaning and has a more familiar style. However, since abstraction is essential to mathematical reasoning (see Sfard, 2000), it could be argued that being knowledgeable in mathematics necessarily involves understanding the language of mathematics (Greeno, 1989) and being able to talk as mathematicians do (see Lave & Wenger, 1991).

Discourse and stylistic properties of writing shape the reader's perception of the extent to which meaning is preserved. Suppose that students in a class are asked to state why, given certain premises, t is a positive number. Here are two hypothetical students' responses:

Student 1: As shown in [5], $t > x + m$. Therefore, $t > 0$.
Student 2: I saw in Equation 5 that $t > x + m$. That is why I think t is positive.

While the two responses show the same conclusion and provide the same kind of justification, Student 1's response could be judged as reflecting a deeper understanding of mathematics than Student 2's response, simply because its style is the same as the style typically used in mathematics textbooks.

In discussing how writing reflects a student's knowledge of mathematics, Morgan (1998) observes that features of text such as "the presence of algebra" (e.g., the use of letters to denote variable names), "abstractness" (e.g., the absence of references to persons, the use of present tense), "correct" terminology, and the absence of evidence of process (e.g., the absence of a step-by-step description of the reasoning used in solving a problem) may influence teachers' judgments of their students' mathematical skills. Thus, mathematical understanding and use of the conventions of the investigation report genre may be difficult to dissociate.

At the core of the debate around the use of mathematical and informal language is the tension between mathematizing thinking and making mathematics meaningful as forms of mathematical enculturation—the entry into the mathematical community through interaction with others (see Schoenfeld, 1992). Some (e.g., Sfard, 2000; Sfard & Cole, 2003) argue that an excessive emphasis on real-life mathematics and real-life context in the teaching of mathematics takes essence away from the discipline—the ability to deal with abstract ideas and symbols. Others (e.g., Brenner, 1998b; Ladson-Billings, 1995) argue in favor of making mathematics meaningful by connecting students' everyday life experiences to school curriculum.

Negotiating Meaning

Findings from research on the role of social interaction in the classroom (e.g., Leung, 2005) underscore the importance of allowing students to construct meaning through language beyond the simple use of mathematical vocabulary. Unfortunately, normative documents may overemphasize "correct vocabulary" and formal language and dismiss the importance of natural language, thus limiting the linguistic resources students can use to construct mathematical knowledge.

Raiker (2002) investigated the extent to which the characteristics of the spoken language used in mathematical conversations influence the teaching of mathematical concepts and examined the possibility that the *National Numeracy Strategy* (NNS)—Great Britain's official document intended to provide teachers with mathematics standards, course structure, and class activities—may overemphasize "the correct use of mathematical vocabulary." Through discourse analysis of classroom interactions, Raiker was able to observe that both teachers and students tend to ascribe different meanings to technical terms. Raiker also observed that teachers' use of some terms included in *Mathematical Vocabulary*—a document that supplements

NNS and which contains terms that its authors believed can facilitate the learning of certain concepts—appeared to hamper, rather than facilitate, the learning of the targeted concepts.

Barwell (2005) notes that, while ambiguity is a common occurrence in mathematical conversations and can be a valuable resource in the teaching of mathematics (see also Rowland, 2000), strategies used by NNS appear to be based on the premise that mathematical language is always precise. He cites statements from *Mathematical Vocabulary* intended to provide teachers with guidance on the use of mathematical language in the classroom:

- "children need support to move on from 'informal' to 'technical' language in mathematics, and from hearing and speaking new vocabulary to reading and writing;
- "teachers should ascertain the extent of children's mathematical vocabulary and the depth of their understanding." (cited by Barwell, 2005, p. 120).

Thus, the guidelines appear to imply that academic language occurs only in the reading and writing modes, which contradicts current thinking in functional linguistics that certain forms of oral discourse are highly academic (Schleppegrell, 2004).

An important issue in research that examines mathematical conversations is the need for appropriate conceptual frameworks for characterizing social interaction in the classroom. Lampert and Cobb (2003) note that the study of the relation between communication in the classroom and student achievement has been based on loose definitions of communication. What counts as student–student interaction or group discussion may be different across studies. Also, features of discourse measured tend to be too generic and may not address aspects of communication that are specific to mathematics.

An example of a conceptual framework for examining classroom interactions comes from the field of formative assessment. Formative assessment has been characterized as the set of assessment activities intended to support learning—assessment *for* learning—as opposed to those intended to appraise learning—assessment *of* learning (Black, 1993; William 1999a, 1999b). Ruiz-Primo and Furtak (2006, 2007) propose a model for examining informal formative assessment in the classroom—the set of unplanned, unstructured forms of assessment that take place in classroom conversations. They characterize classroom interactions as cycles comprising four steps: the teacher elicits a question; the students respond; the teacher recognizes critical information from the students' responses; and the teacher uses that information to support student learning. These cycles are not a prescribed teaching formula. Rather, they occur naturally in classroom conversation and may be initiated at any of its four steps. Ruiz-Primo and Furtak's results show that informal formative assessment is linked to successful student learning. Students who performed well in assessments embedded in instruction tended to have teachers who completed these cycles more frequently.

Language as a System

Views of language as a system are observed in research that addresses the confluence of languages and language varieties in the construction of mathematical language. *System* refers to the fact that different forms of language (e.g., world languages, the dialects of a given language, everyday language) are governed by rules and conventions. It also denotes choice in an individual's use of a language or a dialect according to social contextual factors (see Coulmas, 2005; Fishman, 1965). *Linguistic diversity* refers to different languages (e.g., English, Haitian-Creole), different dialects of a given language (e.g., Standard English, African American Vernacular English), different forms of a language (e.g., informal language, formal language, academic language), different levels of proficiency in a given language, the condition of being bilingual, and each of the languages of a bilingual individual (e.g., the first language, the second language).

Here I discuss language, dialect, and the condition of being bilingual as instances of linguistic diversity. I also discuss three aspects of language in research that addresses language as a system: code-switching in problem solving, the tension between languages, and the influence of language in the interpretation of mathematical problems.

Language, Dialect, and Bilingualism

English, Spanish, Náhuatl, Swahili, and any other language are rule-governed systems, each consisting of a unique set of arbitrary conventions of sounds, symbols, and a unique set of rules for combining those sounds and symbols in ways that allow communication among its users (see Fasold & Connor-Linton, 2006; Wardhaugh, 2002). The word *arbitrary* stresses the fact that no language is more natural than any others. Calling *tree* a tree in English is as natural as calling it with another word in any other language. Likewise, no language is *better* than others, as each language develops in a way that meets the communicative needs of its users (see Nettle & Romaine, 2002).

Dialects are also rule-governed systems, varieties of a same given language that are, within broad limits, mutually intelligible, and which can be distinguished from one another by virtue of such features as pronunciation, grammar, vocabulary, discourse conventions, and the use of certain sets of idiomatic expressions and colloquialisms (see Crystal, 1997). While the term *dialect* is sometimes used to characterize a variety of a language as corrupted, everybody speaks dialects (Preston, 1993). All dialects of the same language have comparable levels of sophistication and complexity (Farr & Ball, 1999). While some dialects may be more prestigious than others, no dialect is better, as system, than others (Corson, 2001). The most prestigious variety of a language is frequently referred to as the standard dialect of that language, such as *Standard English* (Wardhaugh, 2002).

The notion of system also can be applied to examine bilingualism. A bilingual individual can be thought of as someone who has a language system comprising two languages, the native language and the second language (Bialystok, 2001). The condition of being bilingual is not the addition of two separate languages; rather, those two languages make the bilingual person's language system (see Grosjean, 1989). While political discourse sometimes characterizes bilingualism as a deficit (see Crawford, 2000), no scientific evidence exists that supports such conception (Baker, 2006). Indeed, there are some cognitive advantages that result from being bilingual, including an increased flexibility in the performance of certain cognitive tasks and an increased metalinguistic awareness (Bialystok, 2002).

"Bilingual" is a term that reflects a wide range of degrees of proficiency that an individual may have in two languages. This proficiency may vary considerably not only across that person's two languages but also across language modes (e.g., listening, speaking, reading, and writing) within each language. Thus, individuals who are not proficient in the language of instruction (e.g., English language learners) can be viewed as bilingual students, even if their bilingualism is incipient (see Valdés & Figueroa, 1994).

Code-Switching in Problem Solving

Moschkovich's (2006) study of code-switching in mathematical conversations is a good example of research that addresses language choice (see also Moschkovich, 2002). Code-switching is a term from the field of sociolinguistics, which refers to the alternate use of two languages or two dialects during conversation. Current thinking in sociolinguistics and bilingual development holds that, rather than a deficiency, code-switching is a complex skill that involves sophisticated knowledge of the syntactical structures of an individual's two languages (see Bialystok, 2001; Poplack, 1980).

Moschkovich examined the transcript of a conversation between two Grade 9 bilingual, native Spanish speaking Latina students engaged in solving a problem connecting a linear equation and its graph (Figure 4.1).

An analysis of the transcript revealed that code-switching allowed the students to build arguments efficiently. For example, in explaining why the line's steepness should be lower, one of the students, Marcela, said:

8a. If you change the equation $y = x$ to $y = -0.6x$, how would the line change?

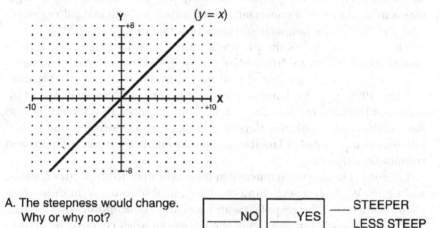

A. The steepness would change.
Why or why not?

_____NO	_____YES	___ STEEPER
		___ LESS STEEP

Figure 4.1 Steepness problem (Moschkovich, 2006).

"Porque fíjate, digamos que este es el suelo.
[Because look, let's say that this is the ground.]
Entonces, si se acerca más, pues es menos steep.
[Then, if it gets closer, then it's less steep.]
... 'cause see this one [referring to the line $y = x$] ... is ...
está entre el medio de la x y de la y. Right?
[is between the x and the y]"

Marcela builds an argument in which the register learned from formal instruction in English is used to refer to a mathematical concept ("steep") and Spanish—Marcela's native language—is used to provide illustrations and emphasize some parts of the argument. A constructive mathematical discussion takes place because the students use their language system without restrictions. A "view of everyday language as unscientific and as contrasted with the precision and specificity of scientific terminology does not do justice to how human beings use language to think and to learn" (Warren, Ballenger, Ogonowski, Rosebery, & Hudicourt-Barnes, 2001, p. 539). Effective learning is more likely to occur when students are allowed to use their linguistic resources in full.

Tension Between Languages

The use of multiple languages in the classroom creates a tension between languages in the sense that different languages have different functions and, therefore, different sets of advantages and resources for constructing mathematical knowledge. For example, in post-apartheid South Africa's multilingual classrooms, the use of students' native languages can be encouraged with the purpose of facilitating the construction of arguments in mathematical conversations. However, a mathematical register is available only in the language of instruction (Setati & Adler, 2001).

This tension parallels the process known in sociolinguistics as *language contact*, which refers to "the outcomes for speakers and their languages when new languages are introduced into a speech community" (Mesthrie & Leap, 1999, p. 248). Language contact occurs as a consequence of an increased interaction "between people from neighbouring territories who have traditionally spoken different languages. But, more frequently, it is initiated by the spread of languages of power and prestige via conquest and colonisation" (p. 248).

Languages interact dynamically in a process that involves power, status, and utility. While code-switching can be a valuable resource in these classrooms, official language policies and teachers' conceptions of language and personal goals can conflict in the classroom practice, creating personal, practical, and contextual dilemmas for teachers. South African teachers who are interested in promoting mathematical conversation among their students through the use of their native languages may also be interested in providing opportunities for them to develop both mathematical language and informal English—a language which they value as a social asset (see Setati, 2002). Since most of the mathematical language is available only in English, it needs to be modeled by the teacher through English. However, modeling English academic language may take valuable time that could be spent in mathematical conversation.

In one of the lessons in multilingual classrooms, Setati (2002) observed that the use of English also produced the dominance of procedural discourse: Students responded in procedural discourse when the teacher asked a conceptual question or remained silent until the teacher asked a procedural question. Setati attributes this finding to the fact that the two forms of discourse have different sets of linguistic and mathematical knowledge demands. While the former can be constructed through simple memorization of the sequence of actions that need to be performed to solve a problem, the latter requires the learner to understand the reasons underlying that sequence of actions. These observations confirm the notion that mathematical intellectual practices are social practices, not simple cognitive routines (O'Connor, 1998). Just the choice of a language can impact, under a given set of circumstances, what is learned and how it is learned.

Adler (1995, 1998) notes that dilemmas involving language choice with regards to supporting mathematics learning are not necessarily problematic and should be seen as "sources of praxis" whose analysis can help teachers to improve their skills. Gutiérrez, Baquedano-Lopez, and Tejeda (1999) focus on the potential for learning of these *hybrid spaces*—as they call the environments in which students are allowed to use multiple forms of language. To them, hybridity and diversity are resources, not challenges; hybrid spaces are environments in which "social, political, material, cognitive, and linguistic [tensions] are sites of rupture, innovation, and change that lead to learning" (p. 287).

Though effective, some strategies intended to address the challenges derived from teaching multilingual classes have their own sets of challenges. Adler uses the term *mathematical language teaching* to refer to teaching in which "language itself, and particularly talk, becomes the object of attention in the mathematics class and a resource in the teaching and learning processes... [and includes] being more explicit about instructions for tasks and more careful in [the] use of mathematical terms and [the] expression of ideas" (Adler, 1999, p. 48). While the use of this approach may be beneficial to all students regardless of their linguistic background, it also may produce an excessive attention to students' mathematical verbalizations and neglect mathematical conversation. Effective teaching in linguistically diverse classrooms requires sophisticated skills among educators, who need to walk a fine line between teaching content and providing the linguistic support needed by students to learn that content.

Language Influences in the Interpretation of Mathematical Problems

The body of research on the linguistic structure of number names and its influence on performance in counting (e.g., Miller & Stigler, 1987; Miura, 1987; Miura, Okamoto, Vlahovic-Stetic, Kim, & Han, 1999) is a good example of research that addresses mathematics learning and teaching with a perspective of language as a system. The transparency with which the naming of numbers reflects the logic behind the base-10 number system varies considerably across cultures (see Saxe, 1988). For example, in English, *eleven, twelve, thirteen, twenty, thirty,* and *forty* do not express clearly the number of units of a decade. In contrast, the names for the same numbers in Chinese and other Asian languages are literally *ten-one, ten-two, ten-three, two tens, three tens,* and *four tens.*

Miura and Okamoto (2003) examine those and other differences between languages and propose that languages provide their users with different sets of supports for the development of mathematics understanding. These differences can affect the ease with which children learn to count and shape the kind of initial exposure they have to mathematics in the context of formal instruction (see Kilpatrick, Swafford, & Findell, 2001). Because languages are governed by different sets of rules, exact equivalence of the same problem across languages is virtually impossible (see Greenfield, 1997). An example of these rules is the use of *ko,* a numeral classifier which is used in Japanese when counting small, round objects. Miura and Okamoto discuss the word problem: *Joe has 6 marbles. He has 2 more than Tom. How many does Tom have?* When translated into Japanese, the problem reads: *Joe has 6 (ko) marbles, 2 (ko) more than Tom. How many (ko) does Tom have?*

Once the referent is established, the noun (marble) can be omitted, but the corresponding numeral classifier (ko) has to be used in the remainder of the problem. Thus, in Japanese, the problem is understood as *Joe has 6 (small, round thing) marbles, 2 (small, round thing) more than Tom. How many (small, round thing) does Tom have?* Miura and Okamoto argue that, in Japanese, numbers in isolation, as in "2 more than Tom" are not an abstract quantity. *Ko* acts as a concept signifier that makes problems more concrete by producing a more vivid representation of the word problem. Thus, the same problem may not pose the same set of challenges in different languages.

FORMAL VIEWS

Language as a Structure

Views of language as a structure are observed in research that examines how mathematical problem understanding is influenced by their linguistic features. *Structure* refers to the organization of text (see Crystal, 1997). The vocabulary (technical and non-technical) and the syntactic complexity of mathematical problems are examined by means of judgmental procedures that focus primarily on grammar constituents (e.g., propositional phrases, verbs, determiners) as units of analysis.

Research that addresses the structural aspects of language faces the challenge that the set of features that make up formal mathematical discourse are so distinctive (e.g., the frequent occurrence of passive voice and nominalization [see Morgan, 1998]), that in some cases mathematical discourse may be confounded with mathematical content. Test writers and test reviewers continuously face the dilemma of using a discourse that is consistent with the discourse of the discipline and using a discourse that does not pose unnecessary reading demands to test takers but may not have the level of abstraction that is perceived as inherent to mathematical reasoning.

Here I discuss two closely related aspects of language as a structure, grammar and semantics. The former refers to the structural complexity of language as a predictor of item difficulty. The latter refers to the relation between wording and semantic relations in mathematics word problems.

Grammar

Research on the linguistic complexity of test items addresses the concern that verbal and reading ability have a significant influence on student performance in mathematics tests (Thurber, Shinn, & Smolkowski, 2002). It is argued, for example, that word problems have characteristics that make them different from other text materials due to their unique style, the abundance of lexical terms, and the scant continuity of ideas across sentences (Davis, 1991; Ferguson & Fairburn, 1985). A great deal of the process of test development has to do with refining the wording of items (see Solano-Flores & Shavelson, 1997). Not surprisingly, even a small change in the wording of an item may affect the semantic structure of test items and, therefore, the way in which students interpret them (De Corte, Verschaffel, & Pauwels, 1990; Shorrocks-Taylor & Hargreaves, 1999).

In their hope to come up with handy tools for examining the linguistic complexity of text in tests, every now and then researchers, practitioners, and even test developers turn to readability formulas created based on counts of such features as sentence length and number of syllables—which are assumed or have been observed to be good predictors of reading level (Gunning, 2003) for a given population and for a specific type of text. Unfortunately, readability formulas have serious limitations derived from the fact that they tend to ignore important factors involved in text comprehension, such as word meaning and the complexity of sentence construction (see Crystal, 1997). To be properly used (but without losing sight of their limitations), they need to be developed from sufficiently large samples of text of the same kind as the text to be examined and with large samples of individuals who are representative of the target population of readers (see Harrison, 1999). Since, by definition, test items consist of small segments of text, any measure of the readability of test items is objectionable (Paul, Nibbelink, & Hoover, 1986).

In an attempt to identify some principles that can guide the process of test development and test adaptation (especially for students with limited proficiency in the language in which tests are administered), researchers have investigated how student performance on science and mathematics test items is affected by linguistic complexity, which is defined in terms of the frequency of technical vocabulary, verb phrases, conditional clauses, relative clauses, and the like (Abedi & Lord, 2001). There is evidence that the linguistic simplification of items can reduce the score gap attributable to language proficiency between English language learners (ELLs) and native English speakers. However, the effects of this form of testing accommodation are moderate (Abedi, Hofstetter, & Lord, 2004; Abedi, Lord, Hofstetter, & Baker, 2000).

Shaftel, Belton-Kocher, Glasnapp, and Poggio (2006) observed that some indicators of the complexity of mathematics items were better predictors of item difficulty for English language learners (ELLs) in earlier school grades than for students in higher school grades. However, they also found that mathematical vocabulary was the only common predictor of item difficulty across grades. Their findings confirm the notion that technical terms pose serious linguistic challenges to students in their attempt to learn or demonstrate conceptual understanding. These findings also speak to the elusive nature of language, whose structural complexity does not account entirely for problem difficulty.

The linguistic properties of test items interact with the contextual information they provide and the students' own experiences. In mathematics word problems, contextual information used with the intent to make them meaningful may be interpreted by test takers in many unexpected ways. Take as an example the Lunch Money item (National Assessment of Educational Progress, 1996), whose linguistic features we (e.g., Solano-Flores & Trumbull, 2003) have examined extensively:

> Sam can purchase his lunch at school. Each day he wants to have juice that costs 50¢, a sandwich that costs 90¢, and fruit that costs 35¢. His mother has only $1.00 bills. What is the least number of $1.00 bills that his mother should give him so he will have enough money to buy lunch for 5 days?

Among the many potential linguistic challenges identified in this item is that "only $1.00 bills" could be interpreted by students in three ways: as restricting the number of dollar denominations (as in *His mother has only dollar bills*), as restricting the number of dollar bills (as in *His mother has only dollars*), and as restricting the amount of money (as in *His mother has only one dollar*). We also observed that, in their responses to this item, some students living in poverty used survival strategies (e.g., giving up on the sandwich, suggesting that Sam ask his mother to give him more money) rather than mathematical strategies such as adding the costs of the sandwich, the juice, and the fruit, then multiplying the result by 5, and then rounding the result to the next higher integer. Linguistic challenges like this cannot be detected unless a careful process of review is used that combines strategies like having the students read the items aloud, interviewing them about their interpretations of the items, and examining their written responses.

Other formal approaches for examining the linguistic complexity of items focus on the syntactic structure of sentences. For example, by using a combination of graph theory (see Harary, 1969) and structural linguistics-based sentence parsing procedures (see van Gelderen, 2000; Veit, 1999), it is possible to detect unnecessary complexity in the structure of sentences (Solano-Flores, Trumbull, & Kwon, 2003). This complexity is reflected by properties such as the number of levels and branches and the number and types of nodes in the graph that represents the structure of a sentence. Figure 4.2 shows two sentences in items from the National Assessment of Educational Progress (1996) public release with different syntactic complexities. One of the sentences is from the Lunch Money item discussed

A measurement of 60 inches is equal to how many feet?

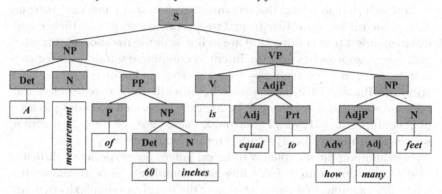

What is the least number of $1.00 bills that his mother should give him so he will have enough money to buy lunch for 5 days?

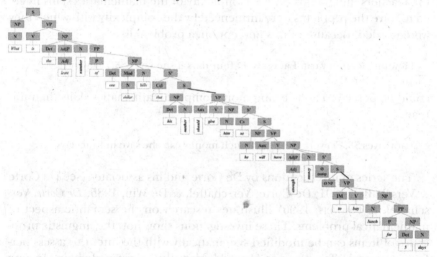

Figure 4.2 Syntactical structure of two sentences from NAEP Grade 4 mathematics items. S = sentence; N = noun; V = verb; Det = determiner; P = phrase; Mod = modifier; Adj = adjective; Aux = auxiliar; NP = noun phrase; VP = verb phrase; AP = adverb phrase; PP = prepositional phrase. (Solano-Flores, Trumbull, & Kwon, 2003).

above. Clearly, the second sentence is more complex than the first and is likely to be considerably more challenging to test takers.

Formal and logical approaches for analyzing problems should be used judiciously. They are valuable approaches for developing and reviewing mathematics items only when they are used in combination with verbal protocols, cognitive interviews, and other empirical data collection procedures.

Semantics

Research that addresses the semantic aspect of mathematical problems aims at identifying how their formal properties are related to their cognitive demands. Efforts of this kind are in line with the tradition of using formal, logical approaches with the intent to enumerate solution strategies—and their degree of correctness—for a given type of mathematical problem (Brown & Burton, 1978) and approaches intended to define problem universes in ways that test items can be logically generated according to a set of generation rules (Hively, Patterson, & Page, 1968) or through mapping sentences (Bormuth, 1970) or other representational devices.

In examining the complexity of items, judges may experience difficulty in distinguishing content from linguistic complexity in mathematics test items. For example, in an investigation of the linguistic complexity of mathematics items, we (Kidron & Solano-Flores, 2006) asked teachers to rank a set of word problems according to their level of difficulty. We observed that the teachers' judgments on the complexity of the mathematics skills needed to solve the problems were influenced by the complexity with which they were worded. Because of this perception, a problem like

How much money can Lara save in four days if she saves $3.75 every day?

could be perceived as assessing more complex mathematics skills than if it read:

Lara saves $3.75 every day. How much money can she save in four days?

The series of investigations by De Corte and his associates (see De Corte & Verschaffel, 1991; De Corte, Verschaffel, & De Win, 1985; *De Corte*, Verschaffel, & Pauwels, 1990) illustrates research on the semantic aspect of mathematical problems. These investigations show how the linguistic properties of items can be modified systematically with the intent to assess performance on different types of problems within a given domain. In one of them, De Corte, Verschaffel, and De Win (1985) studied the effect of rewording on the performance of students on three types of mathematics word problems referred to as *change, combine,* and *compare* problems—basic categories from the classification of addition and subtraction problems originally proposed by Riley, Greeno, and Heller (1983). In change problems, an event changes the value of a quantity (e.g., *Joe won 3 marbles. Now he has 5 marbles. How many marbles did Joe have in the beginning?*); in combine problems, two amounts are combined (e.g., *Tom and Ann have 9 nuts altogether. Tom has 3 nuts. How many nuts does Ann have?*); and in compare problems, two amounts are compared (e.g., *Pete has 8 apples. Ann has 3 apples. How many apples does Pete have more than Ann?*).[1]

De Corte and his colleagues gave problems of the three types to Grade 1 and Grade 2 students in two series. In the first series, the problems were not reworded. In the second series, the problems were reworded by making explicit the semantic relations of their components. For example, after rewording, the problem of Joe and the marbles read: *Joe had some marbles. He won 3 more marbles. Now he has 5 marbles. How many marbles did Joe have in the beginning?* The students' scores and the proportions of correct responses on the reworded problems were significantly higher than the students' scores and the proportions of correct responses on the problems that had not been reworded. These results indicate that problems involving the same kind of arithmetic operation can have different degrees of difficulty due to differences in their underlying semantic structures. Word problems can be reworded to make their semantic relations more explicit without affecting their underlying semantic and mathematical structures, thus making it easier for young students to understand those problems and provide solutions (see De Corte & Verschaffel, 1991).

From a broader perspective, these results also show that even a small modification of the wording of a problem may produce a substantial change in its semantic structure, its difficulty, and the way in which students interpret it. It is because of this complex interaction of linguistic features that De Corte and his colleagues' findings might not be generalizable to more complex problems. As can be seen from comparing the original version and the reworded version of Joe and the marbles, an increased amount of text and an increased set of reading demands may be the price of making the semantic relations among the components of word problems more explicit. This increase in reading demands may be even greater for problems that are substantially more complex.

Language as a Factor

Views of language as a factor are observed in research that uses categories of linguistic groups and treatment conditions, mainly in the context of assessment and, more specifically, in large-scale testing. *Factor* refers to the fact that language is seen as something that needs to be controlled or accounted for in order to obtain accurate measures of mathematics achievement. Score differences between groups are critical to devising strategies intended to reduce the effect of language as an extraneous variable (e.g., Abedi & Lord, 2001; Abedi, Hofstetter, Baker, & Lord, 2001).

Views of language as a factor also are observed in research aimed at detecting and minimizing item bias, systematic performance differences that are attributable to group membership, not the construct being measured (see Schmeiser & Welch, 2006). Approaches derived from item response theory (see van der Linden & Hambleton, 1997; Yen & Fitzpatrick, 2006) are common in research on testing across cultural groups (e.g., van de Vijver & Tanzer, 1997) and testing of groups in different languages (Cook & Schmitt-Cascallar, 2005; Ercikan, 2002). The use of groups is "a statistical device, used because potential bias is uncovered by aggregating evidence across test takers within such groups" (Camilli, 2006, p. 228).

Here I discuss three aspects of research that addressess language as a factor: classifying individuals into population group categories and assigning them to testing conditions, and incorporating language variation into research designs.

Population Group Categories

Language can be addressed by referring to the characteristics of the populations who speak (or do not speak) a given language rather than the characteristics of that language.[2] Individuals are classified into broad linguistic group categories (e.g., English speakers, French speakers), according to a small number of levels of proficiency in a language (e.g., "limited English proficient," "English proficient") or according to a small number of categories intended to describe histories of language development (e.g., "monolingual," "native English speaker").

Examples of research that use population groups with the intent to address language come from the field of international test comparisons such as TIMSS (Trends in Mathematics and Science Study) and PISA (Programme of International Student Assessment), in which different linguistic groups are tested with the same sets of items. The use of item response theory (a psychometric theory of scaling; see van der Linden & Hambleton, 1997) allows detection of biased items—items that are said to function differently across linguistic groups because the performance of the two linguistic groups is not equivalent after controlling for the difference in the overall ability measured (see Allalouf, 2003; Camilli & Shepard, 1994; Hambleton, 2005; Sireci & Allalouf, 2003). There is evidence that differential item functioning can result from very subtle ways in which tests are translated and which affect the equivalence of items across languages. Even the way in which a single word is translated may influence this differential functioning (Ercikan, 1998). Detecting and correcting the origin of this differential functioning may require the use of cognitive interviews with students from the target populations (Ercikan, 2002).

Important differences between international comparisons and the testing of linguistic minorities can be noticed in the ways in which the characteristics of persons are used to group them into linguistic group categories. In the case of international comparisons, linguistic groups are distinguished naturally as a result of their nationalities and their languages of instruction.[3] By contrast, official definitions of "limited English proficient,"

"(25) LIMITED ENGLISH PROFICIENT.—The term 'limited English proficient', when used with respect to an individual, means an individual—
"(A) who is aged 3 through 21;
"(B) who is enrolled or preparing to enroll in an elementary school or secondary school;
"(C)(i) who was not born in the United States or whose native language is a language other than English;
"(ii)(I) who is a Native American or Alaska Native, or a native resident of the outlying areas; and
"(II) who comes from an environment where a language other than English has had a significant impact on the individual's level of English language proficiency; or
"(iii) who is migratory, whose native language is a language other than English, and who comes from' an environment where a language other than English is dominant; and
"(D) whose difficulties in speaking, reading, writing, or understanding the English language may be sufficient to deny the individual—
"(i) the ability to meet the State's proficient level of achievement on State assessments described in section 1111(b)(3);
"(ii) the ability to successfully achieve in classrooms where the language of instruction is English; or
"(iii) the opportunity to participate fully in society.

Figure 4.3   The No Child Left Behind definition of English language learner (No Child Left Behind Act of 2001).

like that shown in Figure 4.3, have many possible interpretations. While this definition acknowledges the multi-faceted nature of language (i.e., it includes the ability to speak, read, write, and comprehend English) and the consequences of not being proficient in English, it does not provide objective criteria for making sound classification decisions on who should be included in the "English language learner" category and who should not. As a consequence of this vagueness, this definition of ELL is likely to be operationalized based on proxy, demographic variables.[4]

An additional problem of this way of defining a linguistic group has to do with its comparability across states. Since different states use different tests to measure English proficiency (National Clearinghouse for English Language Acquisition and Language Instruction Educational Programs, 2006), the category ELL does not have the same meaning, which poses some problems of comparability and equivalence of measures of language proficiency.

Serious limitations of language proficiency measures make one wonder about the extent to which they should be used to make instructional or testing decisions about students. The first limitation has to do with the fact that "language proficiency" is a complex construct that is highly dependent on context. A person can be proficient in a second language for some contexts, and not others (e.g., Bachman, 1990; Canale, 1983; De Avila, 1988;

Grosjean, 1985; Hymes, 1972; MacSwan, 2000). What counts as communicative competence in a given context may not count as communicative competence in another context (see Romaine, 1995; Trumbull & Farr, 2005). In addition, different tests of language proficiency emphasize different language skills (e.g., García, McKoon, & August, 2006). As a consequence, a measure of language proficiency based on a given test may be generalizable only to the set of situations and tasks that are similar to the situations and tasks used by that test.

The second limitation is that, due to their different migration histories, formal education background, and many other reasons, each bilingual individual has a unique pattern of language dominance across the four language modes (listening, speaking, reading, writing) (see Baker, 2006; Bialystok, 2001; Durán, 1989; Solano-Flores & Trumbull, 2008). As a consequence, a measure of language proficiency may give inaccurate information about an individual's actual competencies in a language.

Testing Conditions

An important aspect of the research that addresses language as a factor consists of assigning individuals to different testing conditions such as the language used in a test (e.g., English, Spanish), the language mode in which the test is given to students (e.g., orally, in printed form), or the language mode in which students provide their responses to that test (verbal responses, written responses). A testing condition also may be produced by modifying the linguistic properties of a test with the intent to reduce its linguistic demands (e.g., simplifying the wording of items, including glossaries with word-to-word translations) or from modifying properties of the test that are not related to language but which are thought to be relevant to cognitively processing language (e.g., administering a test with no completion time limit).

A body of research on the use of testing accommodations for ELLs has explored a wide variety of testing conditions intended to reduce the linguistic demands of mathematics and science test items (see reviews by Abedi, Hofstetter, & Lord, 2004 and by Sireci, Li, & Scarpati, 2003). ELL students who receive and who do not receive a form of accommodation are compared to non-ELL students as to their test scores. If the accommodation is effective, then its impact should be reflected as a reduction in the score differences between ELL and non-ELL students; also, the test scores should be higher for ELLs who received the accommodation than for ELLs who did not receive the accommodation. Also, non-ELL students who receive the accommodation are compared with non-ELL students who do not receive the accommodation. If the accommodation truly operates on the linguistic demands of tests, then the scores obtained by non-ELL students with and without the accommodation should not differ substantially.

Not surprisingly, language-related accommodations appear to be more effective than accommodations unrelated to language in reducing the score gap between ELL and non-ELL students (Abedi, 2002; Abedi & Hejri, 2004). More specifically, the linguistic simplification of items appears to be the most effective form of accommodation, although the score differences between ELL and non ELL students are moderate (Abedi et al., 2000). In addition, this form of accommodation does not benefit non-ELL students, which indicates that the linguistic simplification of items is not a threat to the comparability of scores of ELL and non-ELL students (Rivera & Stansfield, 2004).

The effectiveness of testing accommodations may be limited by the lack of fidelity with which they are implemented. For example, while the literature reports with some level of detail the procedures used to create and provide accommodations, the individuals in charge of providing accommodations may not have the qualifications needed (e.g., translation skills, sensitivity to subtle but important dialect variation, knowledge of the mathematical register in the ELL student's native language) to provide them properly. To complicate matters, states vary tremendously as to the types of ELL testing accommodations they use and the kinds of provisions they have for their implementation (see Rivera, Collum, Willner, & Sia, 2006). Thus, a treatment condition such as "test in the student's native language" may mean many different things.

Creating categories of language proficiency and assigning individuals to testing conditions may be appropriate. What is not appropriate is to overestimate the accuracy of measures of language proficiency or to underestimate the conditions that hamper proper implementation of testing conditions (see Solano-Flores, 2008).

Language Variation and Research Design

Assessing measurement error due to language factors rather than pretending that language variation can be controlled for by using a few (and sometimes dubious) categories of language proficiency may be a more effective approach to addressing language in testing. Guided by these reasonings, we (Solano-Flores & Li, 2006) have examined language and dialect as sources of measurement error. We have used a design in which ELL students are tested with the same set of mathematics items in two languages. Rather than testing students in bilingual formats, the intent of this design is to examine score variation across languages and to determine how many items are needed to obtain dependable measures of academic achievement when the students are tested in English and when they are tested in their native language.

By using generalizability theory—a psychometric theory of measurement error developed as an extension of analysis of variance (Brennan, 1992;

Cronbach, Gleser, Nanda, & Rajaratnam, 1972; Shavelson & Webb, 1991)—
we (Solano-Flores & Li, 2006) have been able to identify the amount of score
variation due to the interaction of student, item, rater, and language in open-
ended mathematics tests. Our results show that the performance of ELLs in
mathematics tests is unstable both across items and across languages.

More specifically, our results indicate that, in addition to individual
differences in the mathematical skills measured, each ELL student has a
unique set of strengths and weaknesses in both his/her first language and
his/her second language; also, in addition to its intrinsic cognitive and academic demands, each item poses a specific set of linguistic demands in each
language. Our results also indicate that, due to language variation, localities
may vary tremendously as to both the language that is more appropriate to
use to test ELLs and the number of items needed to obtain dependable
scores. This variation occurs even among ELLs classified within the same
level of English proficiency.

From a more general perspective, our results show that it is possible to use
psychometric models that are consistent with the notion of language variation and with the notion that each bilingual individual has a unique pattern
of language dominance. Notice how, in a design like the one described above,
the effect of language proficiency on performance in tests is not addressed by
comparing ELL students with their mainstream counterparts.

CONCLUDING COMMENTS

In this chapter, I have presented a conceptual framework for examining
research on language and mathematics education. According to this framework, there are four ways to conceptualize language: as a process, as a system, as a structure, and as a factor. Views of language as a process and as a
system can be called functional, because they focus on the role of language
as a phenomenon that influences mathematical communication and the
development of mathematical knowledge. Views of language as a structure
and as a factor can be called formal, because they focus on patterns of
linguistic features of mathematical problems or types of linguistic groups.
No view of language is better than the others; each emphasizes a particular
aspect of language.

My review of the literature shows that research on teaching and learning
(including classroom informal assessment) uses principally functional views
of language, and research and practice in large-scale testing uses principally
formal views of language. This divide makes it difficult to address the relationship between language and mathematics across studies of classrooms
and large-scale testing and confirms the need for multidisciplinary work on

this topic (see Lee, 1999, 2002; Lee & Fradd, 1998; Pellegrino, Chudowsky, & Glaser, 2001).

A review of the translations of mathematics test items from the PISA and TIMSS international test comparisons shows the value of such multidisciplinary work. With proper facilitation, multidisciplinary teams of translation reviewers were able to identify and code test translation error with an unprecedented level of detail (Solano-Flores, Contreras-Niño, & Backhoff-Escudero, 2005). Test translation error is not necessarily a consequence of a poor translation; rather, it is mainly a consequence of the fact that languages encode meaning in different ways (Solano-Flores, Backhoff, & Contreras-Niño, 2009). Translation error is multidimensional. For example, a grammatical error in a translated item can also be a semantic error that alters the original intended meaning of the item in the source language. Because of this multidimensionality, the proper detection of test translation error is more likely to occur when multidisciplinary teams of reviewers that include teachers, mathematicians, curriculum experts, psychometricians, translators, and linguists discuss the potential linguistic challenges of translated items from different perspectives.

Future research work should pay more attention to dialect diversity in mathematics education. Practices in the field of mathematics teaching are influenced by inaccurate assumptions about dialect. While the devastating consequences of devaluing non-standard dialects in the classroom have been discussed extensively (Brisk, 2006; Delpit, 1995; Wolfram, Adger, & Christian, 1999), more research is needed to determine effective ways to train teachers to create respectful, inclusive learning environments that make mathematical knowledge accessible to all students.

Moschkovich (2007) observes that educators tend to base their views about language and learners on vocabulary, on the multiple meanings of words, or on discourse. These views are associated with the views they have about bilingual individuals, respectively as deficient, as facing more difficulties than monolingual students in learning mathematical register and dealing with multiple meanings, and as individuals whose competencies and resources may be comparable to the competencies and resources of mainstream students. According to Moschkovich, only when they have a view of language as discourse are teachers capable of viewing instruction as a means for uncovering bilingual students' competencies and building mathematical knowledge upon those competencies.

Lauren Young (personal communication) has suggested the notion of *mathematical language knowledge* as a form of knowledge that professional development should address along with content mathematical knowledge and pedagogical mathematical knowledge. The contribution of this notion may be worth exploring because it appears to address, in its right dimension, the relevance of language in mathematics teaching and learning. Developing a sophisticated view of language is unlikely to occur simply from being provided with certain language principles. For example, properly addressing the tension between languages in multilingual classrooms takes sophisticated teaching skills, accurate knowledge of language issues, and a favorable attitude towards linguistic diversity.

Although conventional professional development activities can change individuals' stated beliefs about language and its relationship to the teaching of disciplinary knowledge, those changes are not necessarily reflected in teachers' practices (Lee, Hart, Cuevas, & Enders, 2004). A deep transformation of teacher's practices takes place only when there is opportunity for reflection and insight and extensive, continuous support from colleagues and facilitators (Lee, 2004, 2005). In addition to the multiple facets of language and its complex relation to mathematics, the role of culture in the mental representation of mathematics (Stigler & Baranes, 1988–1989) and the social dimension of mathematical thinking (Moses, 2001) need to be considered in order to accurately specify the domain of mathematical language knowledge. This kind of knowledge involves attitudes, beliefs, and thinking and practice that go far beyond basic language principles.

As with research on teaching, future research on mathematics assessment work should pay more attention to dialect diversity. The relevance of dialect as a fairness and validity issue in testing may have been underestimated in the past due to the fact that dialect differences are more subtle and less obvious than language and language proficiency level (see Freedle, 2003). In addition, because of the lack of appropriate knowledge of language, nonstandard dialects may not be properly considered in testing practices. The use of standard English is frequently invoked by test developers as a proof that dialect variation is properly addressed. The underlying assumption is that tests are fair if they are written in standard English because standard English is the dialect that everybody understands. However, current thinking in the field of sociolinguistics recognizes that standard dialect is, in reality, the dialect of the segment of the population of a society that has social and economic power (Halliday, 1978). The linguistic features of a test written in standard English reflect the totality of the features of the dialects used by the privileged segment of the society but only a portion of the features of dialects used by other groups (Solano-Flores, 2006).

There is evidence that subtle but important issues of dialect can be detected and properly addressed if teachers are allowed to participate in the process of test review by discussing items at length and adapting the linguistic features of mathematics tests to the characteristics of the language (either English or their ELL's first language) used in their communities (Solano-Flores, Speroni, & Sexton, 2005). This finding underscores the importance of a simple but frequently ignored fact—that language, as a social phenomenon, can be properly addressed only through social participation.

All these facts speak to the need for a deeper understanding of language issues among researchers, practitioners, and policy makers. If we take seriously the goal of making mathematics accessible to all students, and if mathematics teaching and assessment are to be a coordinated effort, then we need to have clarity about the possibilities and limitations of each view of language across the different aspects of mathematics education. Hopefully, this chapter has contributed to addressing this need.

NOTES

1. The text of the three word problems used as examples is taken verbatim from De Corte, Verschaffel, and De Win (1985).
2. The linguistic group categories used in research and practice in mathematics assessment are determined based on tests of language proficiency or language development whose construction may be based on views of language as a system or as a process. However, this discussion is about how language is addressed in research on mathematics assessment, not how language is measured to classify students according to language proficiency.
3. In some cases, countries participating in international comparisons use two language versions to test different linguistic groups within them (e.g., O'Connor & Malak, 2000).
4. Criteria for determining when a student is no longer an ELL also may be based on erroneous assumptions about language development. Such is the case of the time of schooling as a criterion for deciding when an ELL should be assumed to be able to take tests in English. In spite of the debate over the distinction proposed by Cummins in the early 1980s between the concept of basic interpersonal communicative skills—the skills involved in conversational fluency—and cognitive academic language proficiency—the linguistic proficiency needed to succeed academically—(see Cummins, 2003; MacSwan & Rolstad, 2003; Rivera, 1984), there is consensus among specialists that it is not reasonable to expect that adequate measures of academic achievement can be obtained for ELLs after a short period of immersion in a second language (Guerrero, 2004; Hakuta, 2001; Hakuta, Butler, & Witt, 2000). As discussed in the chapter, academic language involves much more than vocabulary—it also involves skills such as negotiating meaning, constructing an argument, or expressing disagreement (Echevarria & Short, 2002; Scarcella, 2003).

REFERENCES

Abedi, J. (2002). Standardized achievement tests and English language learners: Psychometric issues. *Educational Assessment, 8*(3), 231–257.

Abedi, J., & Hejri, F. (2004). Accommodations for students with limited English proficiency in the National Assessment of Educational Progress. *Applied Measurement in Education, 17*(4), 371–392.

Abedi, J., & Lord, C. (2001). The language factor in mathematics tests. *Applied Measurement in Education, 14*(3), 219–234.

Abedi, J., Hofstetter, C. H., & Lord, C. (2004). Assessment accommodations for English language learners: Implications for policy-based empirical research. *Review of Educational Research, 74*(1), 1–28.

Abedi, J., Hofstetter, C., Baker, E., & Lord, C. (2001). *NAEP math performance and test accommodations: Interactions with student language background.* CSE Technical Report No. 536. Center for the Study of Evaluation, National Center for Research on Evaluation, Standards, and Student Testing.

Abedi, J., Lord, C., Hofstetter, C., & Baker, E. (2000). Impact of accommodation strategies on English language learners' test performance. *Educational Measurement: Issues and Practice, 19*(3), 16–26.

Adler, J. (1995). Dilemmas and a paradox—secondary mathematics teachers' knowledge of their teaching in multilingual classrooms. *Teaching and Teacher Education, 11*(3), 263–274.

Adler, J. (1998). A language of teaching dilemmas: Unlocking the complex multilingual secondary mathematics classroom. *For the Learning of Mathematics, 18*(1), 24–33.

Adler, J. (1999). The dilemma of transparency: Seeing and seeing through talk in the mathematics classroom. *Journal of Research in Mathematics Education, 30*(1), 47–64.

Allalouf, A. (2003). Revising tanslated dfferential iem fnctioning iems as a tool for improving cross-lingual assessment. *Applied Measurement in Education, 16*(1), 55–73.

Bachman, L. F. (1990). *Fundamental considerations in language testing.* Oxford: Oxford University Press.

Baker, C. (2006). *Foundations of bilingual education and bilingualism*(4th ed.). Clevedon, UK: Multilingual Matters.

Barwell, R. (2005). Ambiguity in the mathematics classroom. *Language and Education, 19*(2), 118–126.

Bialystok, E. (2001). *Bilingualism in development.* Cambridge, UK: Cambridge University Press.

Bialystok, E. (2002). Cognitive processes of L2 users. In V. J. Cook (Ed.), *Portraits of the L2 user* (pp. 145–165). Buffalo, NY: Multilingual Matters, Ltd.

Black, P. (1993). Formative and summative assessment by teachers. *Studies in Science Education, 21*, 49–97.

Bormuth, J. R. (1970). *On the theory of achievement test items.* Chicago: University of Chicago Press.

Brennan, R. L. (1992). *Elements of generalizability theory.* Iowa City, IA: The American College Testing Program.

Brenner, M. E. (1994). A communication framework for mathematics: Exemplary instruction for culturally and linguistically diverse students. In D. McLeod (Ed.), *Language and learning: Educating linguistically diverse students* (pp. 233–267). Albany, NY: State University of New York.

Brenner, M. E. (1998a). Development of mathematical communication in problem solving groups by language minority students. *Bilingual Research Journal, 22*, 103–128.

Brenner, M. E. (1998b). Meaning and money. *Educational Studies in Mathematics, 36,* 123–155.

Brisk, M. E. (2006). *Bilingual education: From compensatory to quality schooling* (2nd ed.). Mahwah, NJ: Lawrence Erlbaum Associates, Publishers.

Brown, G., & Yule, G. (1983). *Discourse analysis.* Cambridge, UK: Cambridge University Press.

Brown, J. S., & Burton, R. R. (1978). Diagnostic models for procedural bugs in basic mathematical skills. *Cognitive Science, (2)2,* 71–192.

Burkhardt, H., & Schoenfeld, A. H. (2003). Improving educational research: Toward a more useful, more influential, and better funded enterprise. *Educational Researcher, 32*(9), 3–14.

Camilli, G. (2006). Test fairness. In R. L. Brennan (Ed.), *Educational measurement* (4th ed.) (pp. 221–256). Westport, CT: American Council on Education and Praeger Publishers.

Camilli, G., & Shepard, L. A. (1994). *Methods for identifying biased test items.* Thousand Oaks, CA: SAGE Publications.

Canale, M. (1983). From communicative competence to communicative language pedagogy. In J. C. Richards & R. Schmidt (Eds.), *Language and communication* (pp. 2–27). London: Longman.

Cook, L. L., & Schmitt-Cascallar, A. P. (2005). Establishing score comparability for tests given in different languages. In R. K. Hambleton, P. Merenda, & C. D. Spielberger (Eds.), *Adapting educational and psychological tests for cross-cultural assessment* (pp. 139–169). Mahwah, NJ: Erlbaum.

Corson, D. (2001). *Language diversity and education.* Mahwah, NJ: Lawrence Erlbaum Associates, Publishers.

Coulmas, F. (2005). *Sociolinguistics: The study of speakers' choices.* Cambridge, UK: Cambridge University Press.

Crawford, J. (2000). *At war with diversity: US language policy in an age of anxiety.* Clevedon, UK: Multilingual Matters.

Cronbach, L. J., Gleser, G. C., Nanda, H., & Rajaratnam, N. (1972). *The dependability of behavioral measurements.* New York: Wiley.

Crystal, D. (1997). *The Cambridge encyclopedia of language* (2nd ed.). Cambridge, UK: Cambridge University Press.

Cummins, J. (2003). BICS and CALP: Origins and rationale for the distinction. In C. B. Paulston & G. R. Tucker (Eds.,) *Sociolinguistics: The essential readings* (pp. 322–328). Malden, MA: Blackwell Publishing, Ltd.

Davis, A. (1991). The language of testing. In. K. Durkin & B. Shire (Eds.), *Language in mathematical education: Research and practice* (pp. 40–47). Buckingham, UK: Open University Press.

De Avila, E. A. (1988). Bilingualism, cognitive function, and language minority group membership. In Rodney R. Cocking & Jose P. Mestre (Eds.), *Linguistic and cultural influences on learning mathematics* (pp. 101–121). Hillsdale, NJ: Erlbaum.

De Corte, E., & Verschaffel, L. (1991). Some factors influencing the solution of addition and subtraction word problems. In. K. Durkin & B. Shire (Eds.), *Language in mathematical education: Research and practice* (pp. 17–30). Buckingham, UK: Open University Press.

De Corte, E., Verschaffel, L., & De Win, L. (1985). Influence of rewording verbal problems on children's problem representations and solutions. *Journal of Educational Psychology, 77*(4), 460–470.

De Corte, E., Verschaffel, L., & Pauwels, A. (1990). Influence of the semantic structure of word problems on second graders' eye movements. *Journal of Educational Psychology, 82*, 359–365.

Delpit, L. (1995). *Other people's children: Cultural conflict in the classroom.* New York, NY: New Press.

Durán, R. P. (1989). Testing of linguistic minorities. In R. L. Linn (Ed.), *Educational measurement* (3rd ed.) (pp. 573–587). New York: American Council on Education-Macmillan Publishing Company.

Echevarria, J. & Short, D. (2002). *Using multiple perspectives in observations of diverse classrooms: The sheltered instruction observation protocol (SIOP).* Center for Research on Education, Diversity, and Excellence. http://crede.berkeley.edu/tools/policy/siop/1.3doc2.shtml. Retrieved February 25, 2007.

Ercikan, K. (1998). Translation effects in international assessment. *International Journal of Educational Research, 29*, 543–553.

Ercikan, K. (2002). Disentangling sources of differential item functioning in multi-language assessments. *International Journal of Testing, 2*, 199–215.

Farr, M., & Ball, A. F. (1999). Standard English. In B. Spolsky (Ed.), *Concise encyclopedia of educational linguistics* (pp. 205–208). Oxford, UK: Elsevier.

Fasold, R., & Connor-Linton, J. (2006). *An introduction to language and linguistics.* Cambridge, UK: Cambridge University Press.

Ferguson, A. M., & Fairburn, J. (1985). Language experience for problem solving in mathematics. *Reading Teacher, 38*, 504–507.

Fishman, J. A. (1965). Who speaks what to whom and when? *Linguistique, 2*, 67–88.

Freedle, R. O. (2003). Correcting the SAT's ethnic and social-class bias: A method for re-estimating SAT scores. *Harvard Educational Review, 73*(1), 1–43.

Fuson, K. C. (1991). Children early counting: Saying the number-word sequence, counting objects, and understanding cardinality. In K. Durkin & B. Shire (Eds.), *Language in mathematical education: Research and practice* (pp. 27–39). Buckingham, UK: Open University Press.

Garcia, E. E. (1993). Language, culture, and education. *Review of Research in Education, 19*, 51–98.

García, G. E., McKoon, G., & August, D. (2006). Language and literacy assessment of language-minority students. In D. August & T. Shanahan (Eds.), *Developing literacy in second-language learners: Report of the National Literacy Panel on Language-Minority Children and Youth* (pp. 597–626). Mahwah, NJ: Lawrence Erlbaum Associates, Inc., Publishers.

Greenfield, P. M. (1997). You can't take it with you: Why ability assessments don't cross cultures. *American Psychologist, 52*(10), 1115–1124.

Greeno, J. (1989). For the study of mathematics epistemology. In R. Charles & E. Silver (Eds.), *The teaching and assessing of mathematical problem solving* (pp. 23–31). Reston, VA: National Council of Teachers of Mathematics.

Grosjean, F. (1985). The bilingual as a competent but specific speaker-hearer. *Journal of Multilingual and Multicultural Development, 6*, 467–477.

Grosjean, F. (1989). Neurolinguists, beware! The bilingual is not two monolinguals in one person. *Brain and Language, 36*, 3–15.

Guerrero, M. D. (2004). Acquiring academic English in one year: An unlikely proposition for English language learners. *Urban Education, 39*(2), 172–199.

Gunning, T. (2003). The role of readability in today's classrooms. *Topics in Language Disorders, 23*, 175–189.

Gutiérrez, K. D., Baquedano-Lopez, P., & Tejeda, C. (1999). Rethinking diversity: Hybridity and hybrid language practices in the third space. *Mind, Culture, and Activity, 6*(4), 286–303.

Hakuta, K. (2001). *How long does it take English learners to attain proficiency?* University of California Linguistic Minority Research Institute. Policy Reports. Santa Barbara: Linguistic Minority Research Institute. http://repositories.cdlib.org/lmri/pr/hakuta.

Hakuta, K., Butler, Y. G., & Witt, D. (2000). *How long does it take for English learners to attain proficiency?* Policy Report 2000-1. Santa Barbara, CA: University of California Linguistic Minority Research Institute.

Halliday, M. A. K. (1978). *Language as social semiotic: The social interpretation of language and meaning.* London: Edward Arnold.

Halliday, M. A. K. (1993). Some grammatical problems in scientific English. In M. A. K. Halliday & J. R. Martin (Eds.), Writing science: Literacy and discursive power (pp. 69–85). Pittsburgh, PA: University of Pittsburgh Press.

Hambleton, R. K. (2005). Issues, designs, and technical guidelines for adapting tests into multiple languages and cultures. In R. K. Hambleton, P. Merenda, & C. D. Spielberger (Eds.), *Adapting educational and psychological tests for cross-cultural assessment* (pp. 3–38). Mahwah, NJ: Erlbaum.

Harary, F. (1969). *Graph theory.* Reading, MA: Addison-Wesley Publishing Company, Inc.

Harrison, C. (1999). Readability. In B. Spolsky (Ed.), *Concise encyclopedia of educational linguistics* (pp. 428–431). Oxford, UK: Elsevier.

Hively, W., Patterson, H. L., & Page, S. H. (1968). A "universe-defined" system of arithmetic achievement tests. *Journal of Educational Measurement, 5*(4), 275–290.

Hymes, D. (1972). On communicative competence. In J. B. Price & J. Holmes (Eds.), *Sociolinguistics* (pp. 269–293). Harmondsworth, England: Penguin.

Khisty, L. L. (1995). Making inequality: Issues of language and meaning in mathematics teaching with Hispanic students. In W. G. Secada, E. Fennema, & L. B. Adajian (Eds.), *New directions for equity in mathematics education* (pp. 279–297). Cambridge, UK: Cambridge University Press.

Kidron, Y., & Solano-Flores, G. (2006, April). *Formal and judgmental approaches in the analysis of test item linguistic complexity: A comparative study.* Paper presented at the annual meeting of the American Educational Research Association, San Francisco, California.

Kilpatrick, J., Swafford, J., & Findell, B. (Eds.) and Mathematics Learning Study Committee, National Research Council. (2001). *Adding it up: Helping children learn mathematics.* Washington, DC: National Academy Press.

Ladson-Billings, G. (1995). Making mathematics meaningful in multicultural contexts. In W. G. Secada, E. Fennema, & L. B. Adjian (Eds.), *New directions for equity in mathematics education* (pp. 126–145). Cambridge, UK: Cambridge University Press.

Lampert, M. (1990). When the problem is not the question and the solution is not the answer: Mathematical knowing and teaching. *American Educational Research Journal, 27,* 29–64.

Lampert, M., & Cobb, P. (2003). Communication and language. In J. Kilpatrick, W. G. Martin, & D. Schifter (Eds.), *A research companion to principles and standards for school mathematics* (pp. 237–249). Reston, VA: The National Council of Teachers of Mathematics.

Lave, G., & Wenger, E. (1991). *Situated learning: Legitimate peripheral participation.* Cambridge, UK: Cambridge University Press.

Lee, O. (1999). Equity implications based on the conceptions of science achievement in major reform documents. *Review of Educational Research, 69*(1), 83–115.

Lee, O. (2002). Promoting scientific inquiry with elementary students from diverse cultures and languages. *Review of Research in Education, 26,* 23–69.

Lee, O. (2004). Teacher change in beliefs and practices in science and literacy instruction with English language learners. *Journal of Research in Science Teaching, 41*(1), 65–93.

Lee, O. (2005, April). *Adaptation of an instructional intervention in linguistically, culturally, and socioeconomically diverse elementary schools.* Paper presented at the symposium on "Fidelity of Implementation" at the annual meeting of the American Educational Research Association, Montreal, Canada.

Lee, O., & Fradd, S. H. (1998). Science for all, including students from non-English language backgrounds. *Educational Researcher, 27*(4), 12–21.

Lee, O., Hart, J. E., Cuevas, P., & Enders, C. (2004). Professional development in inquiry-based science for elementary teachers of diverse student groups. *Journal of Research in Science Teaching, 41*(10), 1021–1043.

Leung, C. (2005). Mathematical vocabulary: Fixers of knowledge or points of exploration? *Language and Education, 19*(2), 127–135.

MacSwan, J. (2000). The threshold hypothesis, semilingualism, and other contributions to a deficit view of linguistic minorities. *Hispanic Journal of Behavioral Sciences, 22*(1), 3–45.

MacSwan, J., & Rolstad, K. (2003). Linguistic diversity, schooling, and social class: Rethinking our conception of language proficiency in language minority education. In C. B. Paulston & G. R. Tucker (Eds.), *Sociolinguistics: The essential readings* (pp. 329–340). Malden, MA: Blackwell Publishing, Ltd.

Mesthrie, R., & Leap, W. L. (1999). Language contact 1: Maintenance, shift and death. In R. Mesthrie, J. Swann, A. Deumert, & W. L. Leap (Eds.), *Introducing sociolinguistics* (pp. 248–278). Philadelphia, PA: John Benjamins Publishing Company.

Meyerson, M. J., Ford, M. S., Jones, W. P., & Ward, A. W. (1991). Science vocabulary knowledge of third and fifth grade students. *Science Education, 75*(4), 419–428.

Miller, K. F., & Stigler, J. W. (1987). Counting in Chinese: Cultural variation in a basic cognitive skill. *Cognitive Development, 2,* 279–305.

Mitchell, J. M. (2001). Interactions between natural language and mathematical structures: The case of "wordwalking." *Mathematical Thinking and Learning, 3*(1), 29–52.

Miura, I. T. (1987). Mathematics achievement as a function of language. *Journal of Educational Psychology, 81,* 109–113.

Miura, I. T., & Okamoto, Y. (2003). Language supports for mathematics understanding and performance. In A. J. Baroody & A. Dowker (Eds.), *The development of arithmetic concepts and skills: Constructing adaptive expertise* (pp. 229–242). Mahwah, NJ: Lawrence Erlbaum Associates, Publishers.

Miura, I. T., Okamoto, Y., Vlahovic-Stetic, V., Kim, C. C., & Han, J. H. (1999). Language supports for children's numerical fractions: Understanding of cross-national comparisons. *Journal of Experimental Child Psychology, 74,* 356–365.

Morgan, C. (1998). *Writing mathematically: The discourse investigation.* London, UK: Falmer Press.

Morgan, C. (2005). Words, definitions and concepts in discourses of mathematics, teaching and learning. *Language and Education, 19*(2), 103–117.

Moschkovich, J. N. (2002). A situated and sociocultural perspective on bilingual mathematics learners. *Mathematical Thinking and Learning,* Special issue on Diversity, Equity, and Mathematical Learning, N. Nassir & P. Cobb (Eds.), 4(2&3), 189–212.

Moschkovich, J. (2006). Using two languages when learning mathematics. *Educational Studies in Mathematics, 64*(2), 121–144.

Moschkovich, J. (2007). Bilingual mathematics learners: How views of language, bilingual learners, and mathematical communication affect instruction. In N. S. Nasir & P. Cobb (Eds.), *Improving access to mathematics: Diversity and equity in the classroom* (pp. 89–104). New York: Teachers College Press.

Moses, R. P. (2001). *Radical equations: Math literacy and civil rights.* Boston: Beacon Press.

Nagy, W. (1988). *Teaching vocabulary to improve reading comprehension.* Urbana, IL: National Council of Teachers of English.

Nasir, N. S. (2002). Identity, goals, and learning: Mathematics in cultural practice. *Mathematical Thinking and Learning, 4*(2&3), 213–247.

National Assessment of Educational Progress. (1996). *Mathematics items public release.* Washington, DC: Author.

National Clearinghouse for English Language Acquisition and Language Instruction Educational Programs (2006). *NCELA frequently asked questions.* Revised October 2006. http://www.ncela.gwu.edu/expert/faq/25_tests.htm

National Council of Teachers of Mathematics. (1989). *Curriculum and evaluation standards for school mathematics.* Reston, VA: Author.

National Council of Teachers of Mathematics. (2000). *Principles and standards for school mathematics.* Reston, VA: Author.

Nettle, D., & Romaine, S. (2002). *Vanishing voices: The extinction of the world's languages.* New York: Oxford University Press, Inc.

No Child Left Behind Act of 2001. (2001). Pub. L. No. 107-110, 115 stat. 1961 (2002). http://www.ed.gov/policy/elsec/leg/esea02/107-110.pdf

O'Connor, M. C. (1998). Language socialization in the mathematics classroom: Discourse practices and mathematical thinking. In M. Lampert & M. Blunk (Eds.), *Talking mathematics: Studies of teaching and learning in school* (pp. 17–55). New York: Cambridge University Press.

O'Connor, K. M., & Malak, B. (2000). Translation and cultural adaptation of the TIMSS instruments. In M. O. Martin, K. D.Gregory, & S. E. Stemler (Eds.), *TIMSS 1999 technical report* (pp. 89–100). Chestnut Hill, MA: International Study Center, Boston College.

Paul, D. J., Nibbelink, W. H., & Hoover, H. D. (1986). The effects of adjusting readability on the difficulty of mathematical story problems. *Journal of Research in Mathematical Education, 17*, 163–171.

Pellegrino, J. W., Chudowsky, N., & Glaser, R. (2001). *Knowing what students know: The science and design of educational assessment.* Washington, DC: National Academy Press.

Pimm, C. (1987). *Speaking mathematically: Communication in mathematics classrooms.* London, UK: Routledge & Kegan Paul Ltd.

Poplack, S. (1980). Sometimes I'll start a sentence in Spanish y termino en español: Toward a typology of code-switching. *Linguistics, 18*, 581–618.

Preston, D. R. (Ed.). (1993). *American dialect research.* Philadelphia: John Benjamins.

Raiker, A. (2002). Spoken language and mathematics. *Cambridge Journal of Education, 32*(1), 45–60.

Riley, M. S., Greeno, J. G., & Heller, J. I. (1983). Development of children's problem-solving ability in arithmetic. In H. P. Ginsburg (Ed.), *The development of mathematical thinking* (pp. 153–196). New York: Academic Press.

Rivera, C. (1984). *Language proficiency and academic achievement.* Clevedon, UK: Multilingual Matters.

Rivera, C., & Stansfield, C. W. (2004). The effect of linguistic simplification of science test items on score comparability. *Educational Assessment, 9*(3&4), 79–105.

Rivera, C., Collum, E., Willner, L. N., & Sia, Jr. Jose Ku. (2006). Study 1: An analysis of state assessment policies regarding the accommodation of English language learners. In C. Rivera & E. Collum (Eds.), *State assessment policy and practice for English language learners: A national perspective* (pp. 1–136). Mahwah, NJ: Lawrence Erlbaum Associates.

Romaine, S. (1995). *Bilingualism* (2nd ed.). Malden, MA: Blackwell Publishing.

Rowland, T. (2000). *The pragmatics of mathematics education: Vagueness in mathematical discourse.* London: Falmer Press.

Ruiz-Primo, M. A., & Furtak, E. M. (2006). Informal formative assessment and scientific inquiry: Exploring teachers' practices and student learning. *Educational Assessment, 11*(3 & 4), 205–235.

Ruiz-Primo, M. A., & Furtak, E. M. (2007). Exploring teachers' informal formative assessment practices and students' understanding in the context of scientific inquiry. *Journal of Research Science Teaching, 44*(1), 57–84.

Saxe, G. B. (1988). Linking language with mathematics achievement: Problems and prospects. In R. R. Cocking & J. P. Mestre (Eds.), *Linguistic and cultural influences on learning mathematics* (pp. 47–62). Hillsdale, NJ: Lawrence Erlbaum Associates.

Scarcella, R. C. (2003). *Academic English: A conceptual framework.* Report 2003-1. Santa Barbara, CA: University of California Linguistic Minority Research Institute.

Schmeiser, C. B., & Welch, C. J. (2006). Test development. In R. L. Brennan (Ed.), *Educational measurement* (4th ed.) (pp. 307–353). Westport, CT: American Council on Education and Praeger Publishers.

Schleppegrell, M. J. (2004). *The language of schooling: A functional linguistics perspective.* Mahwah, NJ: Lawrence Erlbaum Associates.

Schoenfeld, A. H. (1992). Learning to think mathematically: Problem solving, metacognition, and sense-making in mathematics. In D. Grouws (Ed.), *Handbook of research on mathematics teaching and learning* (pp. 334–370). New York: MacMillan.

Schoenfeld, A. H. (2004). The math wars. *Educational Policy, 18*(1), 253–286.

Schoenfeld, A. H. (2007). Method. In F. Lester (Ed.), *Handbook of research on mathematics teaching and learning* (2nd ed.) (pp. 69–107). Charlotte, NC: Information Age Publishing.

Setati, M. (2002). Researching mathematics education and language in multilingual South Africa. *The Mathematics Educator, 12*(2), 6–19.

Setati, M., & Adler, J. (2001). Between languages and discourses: Language practices in primary multilingual mathematics classrooms in South Africa. *Educational Studies in Mathematics, 43,* 243–269.

Sfard, A. (2000). On reform movement and the limits of mathematical discourse. *Mathematical Thinking and Learning, 2*(3), 157–189.

Sfard, A., & Cole, M. (2003). *Literate mathematical discourse: What it is and why should we care?* Unpublished manuscript. http://communication.ucsd.edu/lchc/vegas.htm

Shaftel, J., Belton-Kocher, E., Glasnapp, D., & Poggio, G. (2006). The impact of language characteristics in mathematics test items on the performance of English language learners and students with disabilities. *Educational Assessment, 11*(2), 105–126.

Shavelson, R. J., & Webb, N. M. (1991). *Generalizability theory: A primer.* Newbury Park, CA: Sage.

Shorrocks-Taylor, D., & Hargreaves, M. (1999). Making it clear: A review of language issues in testing with special reference to the mathematics tests at key stage 2. *Educational Research, 41*(2), 123–136.

Siegler, R. S., & Jenkins, E. (1989). *How children discover new strategies.* Hillsdale, NJ: Lawrence Erlbaum Associates.

Sireci, S. G., & Allalouf, A. (2003). Appraising item equivalence across multiple languages and cultures. *Language Testing, 20,* 148–166.

Sireci, S. G., Li, S., & Scarpati, S. (2003). *The effects of test accommodation on test performance: A review of the literature* (Research Report 485). Amherst, MA: Center for Educational Assessment.

Sloane, F., & Kelly, A. (2003). Issues in high stakes testing programs. *Theory into Practice, 42*(1), 12–18.

Solano-Flores, G. (2006). Language, dialect, and register: Sociolinguistics and the estimation of measurement error in the testing of English-language learners. *Teachers College Record, 108*(11), 2354–2379.

Solano-Flores, G. (2008). Who is given tests in what language by whom, when, and where? The need for probabilistic views of language in the testing of English language learners. *Educational Researcher, 37*(4), 189–199.

Solano-Flores, G., & Li, M. (2006). The use of generalizability (g) theory in the testing of linguistic minorities. *Educational Measurement: Issues and Practice, 25,* 13–22.

Solano-Flores, G., & Shavelson, R. J. (1997). Development of performance assessments in science: Conceptual, practical and logistical issues. *Educational Measurement: Issues and Practice, 16*(3), 16–25.

Solano-Flores, G., & Trumbull, E. (2003). Examining language in context: The need for new research and practice paradigms in the testing of English-language learners. *Educational Researcher, 32,* 3–13.

Solano-Flores, G., & Trumbull, E. (2008). In what language should English language learners be tested? In R. J. Kopriva (Ed.), *Improving testing for English language learners* (pp. 169–200). New York: Routledge.

Solano-Flores, G., Backhoff, E., & Contreras-Niño, L.A. (2009). Theory of test translation error. *International Journal of Testing, 9,* 78–91.

Solano-Flores, G., Contreras-Niño, L. A., & Backhoff-Escudero, E. (2005, April). *The Mexican translation of TIMSS-95: Test translation lessons from a post-mortem study.* Paper presented at the annual meeting of the National Council on Measurement in Education. Montreal, Quebec.

Solano-Flores, G., Speroni, C., & Sexton, U. (2005, April). *The process of test translation: Advantages and challenges of a socio-linguistic approach.* Paper presented at the annual meeting of the American Educational Research Association. Montreal, Quebec.

Solano-Flores, G., Trumbull, E., & Kwon, M. (2003, April). *The metrics of linguistic complexity and the metrics of student performance in the testing of English language learners.* Symposium paper presented at the 2003 annual meeting of the American Evaluation Research Association, Chicago, IL.

Stigler, J. W., & Baranes, R. (1988–1989). Culture and mathematics learning. *Review of Research in Education, 15,* 253–306.

Thurber, R. S., Shinn, M. R., & Smolkowski, K. (2002). What is measured in mathematics tests? Construct validity of curriculum-based mathematics measures. *School Psychology Review, 31,* 498–513.

Trumbull, E., & Farr, B. (2005). Introduction to language. In E. Trumbull & B. Farr (Eds.), *Language and learning: What teachers need to know* (pp. 1–32). Norwood, MA: Christopher-Gordon.

Valdés, G., & Figueroa, R. A. (1994). *Bilingualism and testing: A special case of bias.* Norwood, NJ: Ablex.

van der Linden, W. J., & Hambleton, R. K. (Eds.). (1997). *Handbook of modern item response theory.* New York: Springer-Verlag.

van de Vijver, F., & Tanzer, N. K. (1997). Bias and equivalence in cross-cultural assessment: An overview. *European Review of Applied Psychology, 47,* 263–279.

van Gelderen, E. (2000). *A grammar of English: Sleeping in mine orchard, a serpent stung me.* Tempe, AZ: Arizona State University. www.public.asu.edu/~gelderen/314text/index.htm.

Veit, R. (1999). *Discovering English grammar.* Needham Heights, MA: Allyn & Bacon.

Vygotsky, L. S. (1986). *Language and thought.* Cambridge, MA: MIT Press. (Originally published in 1936).

Vygotsky, L. S. (1978). *Mind in society.* Cambridge, MA: Harvard University Press.

Wardhaugh, R. (2002). *An introduction to sociolinguistics* (4th ed.). Oxford, UK: Blackwell Publishing.

Warren, B., Ballenger, C., Ogonowski, M., Rosebery, A., & Hudicourt-Barnes, J. (2001). Rethinking diversity in learning science: The logic of everyday languages. *Journal of Research in Science Teaching, 38*(5), 529–552.

Webb. N. M. (1991). Task-related verbal interaction and mathematics learning in small groups. *Journal for Research in Mathematics Education, 22,* 366–389.

Wertsch, J. V. (1985). *Vygotsky and the social formation of mind.* Cambridge, MA: Harvard University Press.

William, D. (1999a). Formative assessment in mathematics. Part 1: Rich questioning. *Equals: Mathematics and Special Educational Needs, 5*(2), 15–18.

William, D. (1999b). Formative assessment in mathematics. Part 2: Feedback. *Equals: Mathematics and Special Educational Needs,* 5(3), 8–11.

Wolfram, W., Adger, C. T., & Christian, D. (1999). *Dialects in schools and communities.* Mahwah, NJ: Lawrence Erlbaum Associates, Publishers.

Yen, W. M., & Fitzpatrick, A. R. (2006). Item response theory. In R. L. Brennan (Ed.), *Educational measurement* (4th ed.) (pp. 111–153). Westport, CT: American Council on Education and Praeger Publishers.

CHAPTER 5

RECOMMENDATIONS FOR RESEARCH ON LANGUAGE AND MATHEMATICS EDUCATION

Judit N. Moschkovich

ABSTRACT

This chapter summarizes the recommendations for future research on language and mathematics education proposed in the four chapters in this volume. These recommendations draw on multiple theoretical and methodological approaches and raise productive questions for further inquiry regarding language and mathematics learning, teaching, and assessment. In the first section I summarize and describe three central recommendations for future research. The second section provides examples of research questions proposed in the four chapters as productive for future studies to pursue. The last section considers implications for instruction and assessment practices for English learners in mathematics classrooms.

Language and Mathematics Education, pages 151–170

151

RECOMMENDATIONS FOR RESEARCH

The four chapters in this volume, although from different theoretical perspectives, come to similar conclusions regarding directions for future research on language and mathematics education. Research studies need to draw on a complex understanding of what "language" is, utilize interdisciplinary theoretical approaches and methods, and consider language issues more broadly as they function in multiple settings that include children from the full range of communities represented in our schools.

Recommendation #1: Recognize the Complexity of Language

As a start, research needs to:

- Recognize the complexity of language use in classrooms and the need to explore language in all its complexity
- Move away from simplified views of language as vocabulary
- Embrace the multimodal and multi-semiotic nature of mathematical activity
- Shift from monolithic views of mathematical discourse and dichotomized views of discourse practices.

Overall, research studies need to be clear and explicit as to how the term language is defined and used. The chapters in this volume document several uses of the term language to refer to the language used in classrooms, in the home and community, by mathematicians, in textbooks, and in test items. It is crucial that researchers clarify how they are using the term language, what set of phenomena they are referring to, and which aspects of these phenomena they are focusing on. Research studies also need to clarify how labels for classrooms, teachers, or learners—such as students who are English learners, bilinguals, or multilinguals—are being used.

However, these multiple referents need not be construed as separate and mutually exclusive categories. For example, one important distinction made in this volume is between language *in* the classroom and language *of* the classroom (see, for example, the distinction made in Chapter 2 by Gutiérrez et al., and Schleppegrell's discussion of Christie's [1991] distinction between instructional and pedagogical registers in Chapter 3). When necessary, researchers may need to distinguish between *language in* the mathematics classroom—mathematical discourse practices such as argumentation, proof, justification, defining, and so on—and *language of* the mathematics classroom, more general classroom discourse practices and ways in which language and discourse are regulated by teachers and students. In other instances, *language in* the mathematics classroom and *language of* the mathematics classroom may be blurred as distinctions (see Chapter 2 for examples). Similarly, students who are bilingual may also be learning English, depending on how these terms are defined.

We recommend moving away from oversimplified views of language as vocabulary. Research needs to leave behind a legacy of overemphasis on correct vocabulary and formal language that would limit the linguistic resources teachers and students can use in the classroom to learn mathematics with understanding. Work on the language of disciplines provides a complex view of mathematical language as not only specialized vocabulary—new words and new meanings for familiar words—but also as extended discourse that includes syntax, organization, register, and discourse practices. Studies need to move beyond interpretations of the mathematics register as merely a set of words and phrases that are particular to mathematics. The mathematics register includes styles of meaning, modes of argument, and mathematical practices (Moschkovich, 2007c) and has several dimensions such as the concepts involved, how mathematical discourse positions students, and how mathematics texts are organized.

Research needs to move from viewing language as autonomous and instead recognize language as a complex meaning-making system. To embrace the nature of mathematical activity as multimodal and multi-semiotic (Chapters 2 and 3, this volume; O'Halloran, 2005), research will need to expand beyond talk to consider the interaction of the three semiotic systems involved in mathematical discourse—natural language, mathematics symbol systems, and visual displays. In particular, studies will need to examine how artifacts serve as mediators and how mathematical activity is embodied. (Chapter 2, this volume).

Research also needs to make a fundamental shift away from conceiving *mathematical discourse* or *mathematical practices* as uniform. Mathematical discourse is not a single, monolithic, or homogeneous discourse. It is a system that includes multiple forms and ranges over a spectrum of mathematical discourse practices such as academic, workplace, playground, home, and so on. Many more research studies are needed to better understand how mathematical practices and discourses differ depending on the setting, context, and circumstances. In particular, studies need to consider what mathematical knowledge and discourse practices learners use in each of many different settings, what knowledge and discourse practices learners use across settings, and how to make visible the ways that learners reason mathematically across settings. Instead of asking general questions such as "Does language impact mathematical reasoning?" research needs to ask how, when, and under what circumstances are language and mathematical reasoning connected, and consider the multiple ways that language functions in different circumstances and for different aspects of mathematical reasoning.

In documenting mathematical practices across settings, researchers should consider the spectrum of mathematical activity as a continuum rather than reifying the separation between practices outside of school and practices in school. Analyses should consider everyday and scientific discourses as interdependent, dialectical, and related rather than assume they are mutually exclusive. Rather than debating whether an utterance, lesson, or discussion is or is not mathematical discourse, studies should instead explore what practices, inscriptions, and talk mean to the participants and how they use these to accomplish their goals.

Overall, we recommend moving away from dichotomies such as everyday/academic, formal/informal, or in-school/out-of-school. Research needs to stop construing everyday and school mathematical practices as a dichotomous distinction for several reasons. First, a theoretical framing of everyday and academic practices (or spontaneous and scientific concepts) as dichotomous is not consistent with current interpretations of these Vygotskian constructs (e.g., O'Connor, 1999; Vygotsky, 1987). Vygotsky (and other theorists) describe everyday and academic practices as intertwined and dialectically connected. Second, because classroom discourse is a hybrid of academic and everyday discourses, multiple registers co-exist in mathematics classrooms (Moschkovich, 2007c). Most importantly, for supporting the success of students in classrooms, academic discourse needs to build on and link with the language students bring from their home communities. Therefore, everyday practices should not be seen as obstacles to participation in academic mathematical discourse, but as resources to build on in order to engage students in the formal mathematical practices taught in classrooms. For example, the ambiguity and multiplicity of meanings in everyday language should be recognized and treated not as a failure to be mathematically precise but as fundamental to making sense of mathematical meanings and to learning mathematics with understanding.

We may even need to consider that mathematical language may not be as precise as experienced by mathematicians, mathematics instructors, or those of us deeply enamored of the precision we imagine mathematics provides. Ambiguity and vagueness have been reported as common in mathematical conversations and have been documented as resources in teaching and learning mathematics (e.g., Barwell, 2005; Barwell, Leung, Morgan, & Street, 2005; O'Halloran, 2000; Rowland, 1999). Even definitions are not a monolithic mathematical practice, since they are presented differently in lower-level textbooks—as static and absolute facts to be accepted—while in journal articles they are presented as dynamic, evolving, and open to decisions by the mathematician (Morgan, 2004). Neither should textbooks be seen as homogeneous. Higher-level textbooks are more like journal articles in allowing for more uncertainty and evolving meaning than lower-level textbooks (Morgan, 2004), evidence that there are multiple approaches to the issue of precision, even in mathematical texts.

Recommendation #2: Draw on Interdisciplinary Approaches and Methods

The four chapters in this volume also conclude that research on language and mathematics education must be grounded not only in current theoretical perspectives of mathematical thinking, learning, and teaching, but also in current views of language, classroom discourse, bilingualism, second language acquisition, and assessment design. Research needs to:

- Consider interdisciplinary approaches
- Use frameworks for recognizing the mathematical reasoning learners construct in, through, and with language
- Design studies that build on the findings and instruments used in previous relevant research literature
- Consider multiple methods for data collection and analysis.

Since mathematical activity is multimodal and multi-semiotic, and mathematical understanding involves multiple modalities and artifacts— including oral and written language, gestures, the body, inscriptions, and so on—the study of language and mathematics requires interdisciplinary approaches. In this volume we have reported on several approaches—situated cognition, anthropological, cultural historical activity theory, systemic functional linguistics, applied linguistics, ethnomathematics, Goffman's notion of frames, discursive psychology, and embodied knowing—and there are many more that may be relevant. Future studies should use scholarly literature from different relevant fields. It is important to draw on relevant studies, even when these studies are from different content areas. Studies focused on science classrooms and discourse (e.g., Lemke, 1990; Rosebery, Warren, & Conant, 1992; Warren, Ballenger, Ogonowski, Rosebery, & Hudicourt-Barnes, 2001; Warren, Ogonowsky, & Pothier, 2005) may be relevant to research in mathematics classrooms.

In order to focus on the mathematical meanings learners construct rather than the mistakes they make or the obstacles they face, researchers will need to use frameworks for recognizing the mathematical reasoning that learners are constructing in, through, and with language. There are multiple theoretical frameworks available to accomplish this. A few that were mentioned in this volume include systemic functional linguistics (e.g., O'Halloran, 1999; Schleppegrell, 2007), a communication framework for mathematics instruction (Brenner, 1994), a sociocultural perspective on bilingual mathematics learners (Moschkovich, 2002), ethno-mathematical approaches (e.g., D'Ambrosio, 1985, 1991), cultural-historical-activity-theory (Cole & Engestrom, 1993; Engestrom, 1987; Gutiérrez, 2008), and the intersection of psychometrics and sociolinguistics (e.g., Solano-Flores, 2006).

The design of data collection should build on instruments used in previous research literature that is relevant, such as assessments of language proficiency in particular topics in a home language (e.g., whole number operations in Secada, 1991) or assessments of reading proficiency in English that use traditional word problems (e.g., Clarkson & Galbraith, 1992). When designing research studies, researchers should keep in mind that learners from any community participate productively in a variety of roles, responsibilities, communication styles, and mathematical activities that involve hybrid practices (Gutiérrez, Baquedano-Lopez, & Tejeda, 1999). When participants use two languages, it is imperative that both the actual utterances of the participants in one (or more) language(s) as well as the translations be included in presentations and publications reporting on the research.

Methodological possibilities range from ethnographic studies to grammatical analyses to discourse analysis. One particular recommendation is that future research studies focus on more than brief interactional episodes or dialogue fragments and examine how language and ways of talking evolve over longer periods of time—for example, a whole lesson, a set of related lessons, or a unit of study. Whenever possible, studies should document and report not only bilingual students' proficiency in both their first and second languages but also their histories, practices, and experiences with each language across a range of settings and tasks, especially previous instruction in mathematics in each language. This is also important data to collect for teachers working in classrooms with bilingual students.

In particular, recommendations for assessment research include combining formal approaches to the study of test items with other methods and developing methodology for the use of readability formulas that is appropriate for research on language and mathematics. Studies need to include methods such as verbal protocols, cognitive interviews, and other sources of empirical data in the study of test items. These studies also need to address both the potential of readability formulas as tools in educational research concerning mathematics and language and the danger in using these formulas in linguistic contexts for which they were not originally constructed. Lastly, assessment studies need to develop methodologies for the use of readability formulas in research on mathematics and language which address issues such as the appropriate size of samples of text and samples of individuals to ensure those samples are representative of the target forms of test and reader populations.

Recommendation #3: Consider Language Issues in Multiple Settings

The four chapters in this volume conclude that many more studies are needed that focus on students from non-dominant communities—students who speak more than one language, students who speak non-standard varieties of English, immigrant students, and students who bring multiple repertoires of practice to school (Gutiérrez & Rogoff, 2003). Overall, the authors recommend that these studies:

- Avoid using deficit models of learners, their parents, or their communities
- Consider second language learners and speakers of non-standard varieties of English as they engage in mathematical activity in multiple settings
- Address the nature of mathematical activity and discourse in classrooms with students from non-dominant communities.

One way to avoid deficit models is to consider not only the challenges students face but also the resources (e.g., Gonzalez, Andrade, Civil, & Moll, 2001; Moschkovich, 2000) and competencies (e.g., Moschkovich, 2002, 2007a, 2007b) they bring to mathematics classrooms. Another way to avoid deficit models is to move away from comparisons to a norm. For example, comparisons between bilingual and monolingual speakers are not a useful focus in mathematics classrooms (Moschkovich, 2007d, 2009), because they ignore competencies that distinguish fluent bilinguals—such as code-switching (Zentella, 1981)—and miss how bilingual language competence is simply different from monolingual competence (Bialystok, 2001; Cook, 1997). Comparisons between monolingual and bilingual learners, or students from dominant and non-dominant communities, or speakers of standard English and speakers of other varieties, and so on, assume monolingualism, standard English, or living where one was born as the norms for student experiences. Instead of focusing on comparisons to a norm that few students from non-dominant communities fit, studies need to examine student competencies in their own right and explore the complexity of the experiences of students from non-dominant communities as they relate to mathematical reasoning, learning, and instruction.

Research needs to study second language learners and speakers of non-standard varieties of English engaged in mathematical activity in a range of social settings, both in and out of school. In school, studies need to consider not only the challenges students face—for example, in remedial mathematics classes—but also the successes students experience in mathematics/science academies, advanced classes, or with teachers with a track record of success with this student population. Studies may need to consider cultural differences related to communication in classrooms—for example, norms for responding to questions or answering elders. When exploring cultural differences, it is important that studies frame these as repertoires of practice—not individual characteristics (Gutiérrez & Rogoff, 2003)—and document both the regularities and the variation in the repertoires that children from non-dominant communities bring to school (Gutiérrez, 2004).

In particular, research needs to explore in more detail how immigrant students transition from learning mathematics in the home language to learning mathematics in English. These studies will need to distinguish between different types of English learners—for example, students who have developed literacy in their first language, worked at grade level in mathematics in their first language, and those whose whole education has been in English but spent early years learning English (and may have missed out on some instruction); those students whose education has been interrupted, students who have not developed literacy in their first language, or have not had access to adequate mathematics instruction in any language.

Lastly, research needs to address and carefully examine the nature of mathematical discourse in classrooms with students from non-dominant communities. While we need many more examples of "what is possible" in classrooms with students from non-dominant communities, we also need analyses that describe how teachers and students attend explicitly to mathematical content. Studies are needed that examine whether and how classroom discourse centers on tasks with high-level cognitive demand and provides opportunities for conceptual development. We need many more examples of classroom discourse that supports both procedural and conceptual discussions, where teachers and students attend explicitly to concepts, and where students wrestle with important mathematics.

According to a recent review of the research literature in mathematics education (Hiebert & Grouws, 2007), teaching that makes a difference in student achievement and promotes conceptual development in mathematics has two central features: one is that teachers and students attend explicitly to concepts and the other is that students wrestle with important mathematics. Research studies need to provide the details of how instruction in classrooms with these·student populations can simultaneously provide explicit attention to concepts while allowing students to wrestle with these concepts. One way to ensure that students wrestle with important mathematics is for teachers to choose, set, and maintain tasks at high-level cognitive demand (Stein, Smith, Henningsen, & Silver, 2000). However, Hiebert and Grouws (2007) caution against stopping at the analysis of the kind of problems teachers present to students. They emphasize that describing teaching involves close inspection of how students and teachers interact about content "with considerable detail and precision" (p. 393). More studies are needed that show that it is possible for teachers and students in these classrooms to pay explicit attention to concepts in ways other than providing definitions or stating general principles.

Since we cannot expect that any one way of interacting and using language will be optimal for all students, research will need to examine how students in different types of classrooms experience mathematics instruction differently and are offered different opportunities for access to different mathematical discourse practices. We need more studies that consider how opportunities to learn mathematical discourse vary in relation to social class, gender, and ethnicity. For example, studies might explore the frequency of instructional activities that relate to the mathematics content and those that disrupt a focus on the content or examine how the pedagogical register positions students in ways that provide differential access to participation.

It is important for research that focuses on students from non-dominant communities to draw on relevant studies, even when those studies were not focused on the same student population or conducted in mathematics classrooms. For example, studies in South African multilingual mathematics classrooms (e.g., Adler, 2001; Setati, 1998; Setati & Adler, 2001), or with Hawaiian students learning to read (e.g., Au & Jordan, 1981), or with students who speak Haitian Creole in science classrooms (e.g., Ballenger, 2000; Hudicourt-Barnes, 2003; Warren et al., 2001) can contribute to understanding mathematics learning and teaching for other student populations, as long as the differences among contexts are acknowledged.

QUESTIONS FOR FUTURE RESEARCH

The following is a selection of research questions proposed in the four chapters as productive for moving research in new directions, developing theory, and improving practice.

Proposed research questions focusing on *students and learning* include:

- How do learners develop scientific, school-based, or literate mathematical discourses from everyday or colloquial discourses?
- How does movement back and forth between everyday and scientific mathematical discourses facilitate learning?
- How do the social organization of different settings and different forms of mediation support, facilitate, or interfere with learners developing mathematical understanding?
- How does embodied mathematics learning occur in a range of practices?

- How do differences in student participation, positioning of students, and student resistance to teacher evaluations impact how students see themselves as more or less competent in mathematics?
- What are students' experiences learning mathematics in their first and second languages?
- How do different proficiencies in a first language (oral, reading, written, academic English) and previous mathematics instruction in a first language impact students' learning of mathematics in English?

Proposed research questions focusing on *teachers and teaching* include:

- How do teachers use talk to effectively build on students' everyday language as well as developing their academic mathematical language?
- How do teachers provide interaction, scaffolding, and other supports for learning academic mathematical language?
- How do mathematics teachers make judgments about defining terms and allowing students to use informal language in mathematics classrooms?
- How do teachers decide when imprecise or ambiguous language might be pedagogically preferable and when not?
- How can teachers be better prepared to deal with the tensions around language and mathematical content?
- How can mathematics teacher preparation make knowledge of mathematical talk, language, and discourse available to teachers?
- How can mathematics teacher preparation raise teachers' awareness about language, provide teachers with ways to talk explicitly about language, and model ways to respond to students?

Many more studies are needed that describe how students use multiple resources such as two languages, gestures, objects, embodiment, and inscriptions to communicate mathematically. Studies will need to distinguish among multiple modalities (written and oral) as well as between receptive and productive skills. Other important distinctions are between listening and oral comprehension, comprehending and producing oral contributions, and comprehending and producing written text. Research needs to examine classroom language at different levels, for different topics, and in classrooms with different social organization. Studies will need to clarify the role of registers, distinguish the use of different registers, and consider settings where students speak dialects or languages that do not currently provide mathematics registers.

Academic English for mathematical communication is another important topic for research studies to address. Studies need to examine in more detail what, exactly, constitutes competency in academic English for mathematics, in both written and oral modes. If it is the case that academic English is different for different mathematical domains or genres of mathematical texts, then these differences need to be examined in detail.

While learning to read in mathematics is a topic that needs attention, research should consider how students learn to read different mathematical texts, not only textbooks and word problems, but also newspaper articles that use statistics, web materials that require quantitative literacy, and multimedia representations of mathematical concepts. In designing this research it will be important to differentiate between reading textbooks and reading word problems—two different genres in written mathematical discourse—and other genres of mathematical texts. Studies are also needed that examine movement and "intertextuality" between and among different types of texts, spoken and written, among texts such as homework, blackboard diagrams, textbooks, interactions between teacher and students, and interactions among students.

Proposed research questions focusing on *assessment* include:

- What are the relationships between how known and unknown quantities are expressed, the wording of items, and the features of distractors in multiple-choice questions in mathematics tests?
- What is the relationship between mathematics test score variation and language variation?
- How do aspects of linguistic complexity beyond readability—such as the syntactic structure of sentences, underlying semantic structures, or frequency of technical vocabulary, verb phrases, conditional clauses, relative clauses, and so on—affect mathematics tests and accommodations?
- How are mathematics testing accommodations currently implemented with English learners?

Since mathematics tests assess language skills in addition to mathematical understanding, studies of assessment should pay more attention to the relationship between test score variation and language variation and to dialect diversity, especially in terms of its relevance for fairness and validity issues. These issues are not exclusive to English learners and should involve multiple semiotic systems and multiple aspects of language in mathematics—dialect, language proficiency, testing register, item format, notation, and so on. Test development and adaptation need to consider not only readability but also aspects of linguistic complexity such as the syntactic structure of sentences, underlying semantic structures, or frequency of technical vocabulary, verb phrases, conditional clauses, and relative clauses. A functional systemic linguistics perspective may offer useful constructs for such studies. Assessment systems should evaluate the extent to which forms of testing accommodations currently used with English learners are theoretically grounded and supported by current research on language. Since linguistic simplification is only moderately effective as a form of testing accommodation for English learners, research should explore innovative forms of testing accommodation.

IMPLICATIONS FOR INSTRUCTIONAL AND ASSESSMENT PRACTICES FOR ENGLISH LEARNERS

Although language issues are important to consider for practice in *all* mathematics classrooms, we seem most concerned with issues of language as they arise in classrooms with students who are learning English as a second language. Language issues may seem more salient when it is obvious that the teacher and students do not share a common language for instruction. As the population of English learners increases in U.S. public schools,[1] more teachers are concerned with the needs of these students in mathematics classrooms.

Instructional and assessment practices for students who are English learners should be based on the most recent research literature on language and mathematics education that recognizes the complexity of language. Although early studies (e.g., Mestre, 1988; Spanos, Rhodes, Dale, & Crandall, 1988) provide a good introduction to the issues, scholarly work has forged ahead, creating new ways to frame studies theoretically and providing new empirical evidence for claims about language and mathematics.

Although the chapters in this volume did not consider research on policy directly, two issues specific to English learners are crucial to instruction and assessment policies for this student population:

- The label "English learner" as currently used in the U.S. is vague, has different meanings, is not based on objective criteria, does not reflect sound classifications, and is not comparable across states or equivalent across settings. This label is likely to reflect or be used as a proxy for demographic labels rather than as an accurate portrayal of students who are learning English (see, e.g., Gándara & Contreras, 2009).
- Language proficiency is a complex construct that can reflect proficiency in multiple contexts, modes, and academic disciplines. Current measures of language proficiency may not give an accurate picture of an individual's language competence. In particular, we do not have measures or assessments for language proficiency related to competence in mathematics for different ages or mathematical topics.

These two facts can severely limit and confuse discussions of mathematics instruction for English learners. In particular, because of the complexity of language proficiency and the limitations of the label "English learner" as currently implemented, instructional decisions should not be made solely based on the label "English learner." However, research on language and mathematics education for this student population does provide a few clear results to guide *instructional practices* for teaching English learners mathematics:

- Mathematics instruction for English learners should address much more than vocabulary.
- English learners can participate in mathematical discussions as they learn English (Moschkovich, 1999a, 2002, 2007a, 2007b, 2007d).
- Mathematics instruction for students who are learning English should draw on multiple resources available in classrooms—such as objects, inscriptions, and gestures—as well as home languages and mathematical activity outside of school.

Overall, English learners and students from non-dominant communities need access to curricula, instruction, and teachers who have proven to be effective in supporting the academic success of these students. The general characteristics of such environments are that curricula provide "abundant and diverse opportunities for speaking, listening, reading, and writing" and that instruction "encourage students to take risks, construct meaning, and seek reinterpretations of knowledge within compatible social contexts" (Garcia & Gonzalez, 1995, p. 424). Some of the characteristics of teachers who have been documented as being successful with students from non-dominant communities are: (1) a high commitment to students' academic success and to student–home communication, (2) high expectations for all students, (3) the autonomy to change curriculum and instruction to meet the specific needs of their students, and (4) a rejection of models of their students as intellectually disadvantaged. Curriculum policies for English learners in mathematics should follow the guidelines for traditionally underserved students (American Educational Research Association [AERA], 2004a), such as instituting systems that broaden course-taking options and avoiding systems of tracking students that limit their opportunities to learn and delay their exposure to college-preparatory mathematics coursework.

Mathematics instruction for English learners should follow the general recommendations for high quality mathematics instruction: (1) raise cognitive demand in mathematics teaching and learning in both elementary and secondary schools; (2) students should focus on mathematical concepts and connections among those concepts; and (3) teachers should reinforce high cognitive demand and maintain the rigor of mathematical tasks, for example, by encouraging students to explain their problem-solving (AERA, 2006). In particular, recommendations for supporting English learners in developing literacy (AERA, 2004b) include (1) providing structured academic conversation, built around text and subject matter activities to develop vocabulary and comprehension, and (2) providing several years of intensive, high-quality instruction to help students master the vocabulary, comprehension, and oral language skills that will make them fully fluent in speaking, reading, and writing English.

In Chapter 4, Solano-Flores provides two main suggestions for *assessment practices* in mathematics for English learners:

- Test development should consider English learners
- Assessment systems should consider how testing accommodations for English learners are developed and implemented

Test items should be reviewed using strategies such as having students read items aloud, interviewing students about interpretations of items, and examining written responses. Linguists and cultural anthropologists should be included, along with psychometricians, teachers, and content specialists, as critical members of test development teams throughout the entire process of test development rather than just as consultants at the end of it. Since most English learners are tested in English, they should be included among the samples of pilot students at all stages in the process of test development; they provide valuable information for improving the wording of items.

Assessment systems need to include linguistic simplification as an effective form of accommodation. Translations of tests should be very carefully considered since differential item functioning can result and affect the equivalence of items across languages. Translations and adaptations of tests for linguistic or cultural groups for which they were not originally created should be informed by both psychometric approaches (e.g., the analysis of item differential functioning) and judgmental review approaches (e.g., translation review performed by multidisciplinary teams of specialists). Assessment systems should continuously consider and evaluate how testing accommodations for English learners are implemented, rather than assuming that accommodations are interpreted and implemented uniformly.

Teacher Professional Development

Teacher professional development for mathematics instruction with English learners should follow the policy guidelines for teacher professional development in general: It should focus on the subject matter teachers will be teaching, align teachers' learning opportunities with their real work experiences, use actual curriculum materials and assessments, and emphasize observing and analyzing students' understanding of the subject matter (AERA, 2005).

In order to address issues of language, teacher professional development should support teachers in shifting from seeing language merely as a vehicle for assessing students to seeing language as the medium through which learning and teaching occurs. It should also help teachers see a focus on language as an opportunity to hear students' mathematical thinking and design further instruction based on that thinking (Moschkovich, 1999a, 2002). Such programs will need to pay attention to supporting teachers in learning to improve student competence in articulating what they know as well as create respectful learning environments that provide opportunities for all students to learn to communicate mathematically (Moschkovich & Nelson-Barber, 2009).

In particular, the chapters in this volume recommend that teacher professional development programs focusing on mathematics instruction for English learners:

- Address common tensions described in the literature such as between content and language learning goals, multiple national languages in some classrooms, or the regulatory and instructional uses of pedagogical language
- Make the nature of mathematical language practices themselves a point of discussion among teachers and between teachers and their students
- Involve teachers in reflection, discussion, and continuous interaction with colleagues about issues of language in teaching mathematics
- Involve teachers in reviewing tests, discussing test items, and adapting items

CLOSING

The four reviews of research addressing issues of language in mathematics education in this volume were motivated by concerns about the directions of future research on this topic as well as concerns about practice, especially current issues in mathematics instruction for English learners. We could imagine that the "solution" to the "problem" of mathematics instruction for English learners involves a quick fix: new manuals for teachers, a new piece of software, and so on. Unfortunately, such quick fixes run the risk of reifying myths about language, learning, and mathematics. While this volume is aimed at informing future research, it is also fundamentally informed by the reality of schools, classrooms, and practitioners' needs. Our goals are to support future research so that practice can be changed in principled ways that are based on research rather than myths. Our recommendations are motivated by a commitment to improving mathematics learning for all students and the assumption that language is not "the problem." Although we do not expect a quick fix, the collection of studies examined in this volume provides a solid foundation for future research on language and mathematics education. We now have the opportunity to move knowledge forward by addressing fundamental questions about language and learning, while also making a difference for students in our schools. It is our sincere hope that this volume helps in both of these endeavors.

NOTE

1. In 2001, 4.5 million of K–12 students in public schools (9.3%) were labeled as English Learners (Tafoya, 2002). Between 1979 and 2006 the number of school-age children (ages 5–17) who spoke a language other than English at home more than doubled, increasing from 3.8 million, 9% of the population, to 10.8 million, 20% of the population (Planty et al., 2008). The majority of English learners in the U.S. are Latinos/as. In 2006, about 72% of school-age children (ages 15–17) who spoke a language other than English at home spoke Spanish (Planty et al., 2008). In some states the numbers are even greater. For example, in California, 25% (1.5 million) of the children in public school in 2001 were labeled "English learners," and 83% of those children spoke Spanish as their primary language (Tafoya, 2002).

REFERENCES

American Educational Research Association. (2004a). Closing the gap: High achievement for students of color. *Research Points, 2*(3).

American Educational Research Association. (2004b). English language learners: Boosting academic achievement. *Research Points, 2*(1).

American Educational Research Association. (2005). Teaching teachers: Professional development to improve student achievement. *Research Points, 3*(1).

American Educational Research Association. (2006). Do the math: Cognitive demand makes a difference. *Research Points, 4*(2).

Adler, J. (2001). *Teaching mathematics in multilingual classrooms.* Dordrecht, The Netherlands: Kluwer Academic Press.

Au, K., & Jordan, C. (1981). Teaching reading to Hawaiian children: Finding a culturally appropriate solution. In H. Trueba, G. Guthrie, & K. Au (Eds.), *Culture and the bilingual classroom: Studies in classroom ethnography* (pp. 139–152). Rowley, MA: Newbury.

Ballenger, C. (2000). Bilingual in two senses. In Z. F. Beykont (Ed.), *Lifting every voice: Pedagogy and the politics of bilingualism* (pp. 95–112). Cambridge, MA: Harvard Education Publishing Group.

Barwell, R. (2005). Ambiguity in the mathematics classroom. *Language and Education, 19*(2), 117–125.

Barwell, R., Leung, C., Morgan, C., & Street, B. (2005). Applied linguistics and mathematics education: More than words and numbers. *Language and Education, 19*(2), 141–146.

Bialystok, E. (2001). *Bilingualism in development: Language, literacy, and cognition.* New York: Cambridge University Press.

Brenner, M. (1994). A communication framework for mathematics: Exemplary instruction for culturally and linguistically diverse students. In B. McLeod (Ed.), *Language and learning: Educating linguistically diverse students* (pp. 233–268). Albany, NY: SUNY Press.

Christie, F. (1991). First- and second-order registers in education. In E. Ventola (Ed.), *Functional and systemic linguistics* (pp. 235–256). Berlin: Mouton de Gruyter.

Clarkson, P., & Galbraith, P. (1992). Bilingualism and mathematics learning: Another perspective. *Journal for Research in Mathematics Education, 23*(1), 34–44.

Cole, M. & Engestrom, Y. (1993). A cultural historical approach to distributed cognition. In G. Salomon (ed.), *Distributed cognitions: Psychological and educational considerations* (pp. 1–46). Cambridge, UK: Cambridge University Press.

Cook, V. (1997). The consequences of bilingualism for cognitive processing. In A. de Groot & J. Kroll (Eds.), *Tutorials in bilingualism* (pp. 279–299). Mahwah, NJ: Erlbaum.

D'Ambrosio, U. (1985). *Socio-cultural bases for mathematics education.* Campinas, Brazil: UNICAMP.

D'Ambrosio, U. (1991). Ethnomathematics and its place in the history and pedagogy of mathematics. In M. Harris (Ed.), *Schools, mathematics and work* (pp. 15–25). Bristol, PA: Falmer Press.

Engestrom, Y. (1987). *Learning by expanding. An activity theoretical approach to developmental research.* Helsinki, Finland: Orienta-Konsultit Oy.

Gándara, P., & Contreras, F. (2009). *The Latino education crisis: The consequences of failed social policies.* Cambridge, MA: Harvard University Press.

Garcia, E., & Gonzalez, R. (1995). Issues in systemic reform for culturally and linguistically diverse students. *Teachers College Record, 96*(3), 418–431.

González, N., Andrade, R., Civil, M., & Moll, L. C. (2001). Bridging funds of distributed knowledge: Creating zones of practices in mathematics. *Journal of Education for Students Placed at Risk, 6,* 115–132.

Gutiérrez, K. (2004). *Rethinking Education Policy for English Learners.* Washington, DC: Aspen Institute.

Gutiérrez, K. (2008). Developing a Sociocritical Literacy in the Third Space. *Reading Research Quarterly. 43*(2), 148–164.

Gutiérrez, K., & Rogoff, B. (2003). Cultural ways of learning: Individual traits or repertoires of practice. *Educational Researcher, 32*(5), 19–25.

Gutiérrez, K., Baquedano-Lopez, P., & Tejeda, C. (1999). Rethinking diversity: Hybridity and hybrid language practices in the third space. *Mind, Culture, and Activity, 6*(4), 286–303.

Hiebert, J., & Grouws, D. (2007). The effects of classroom mathematics teaching on students' learning. In F. Lester (Ed.), *Second handbook of research on mathematics teaching and learning* (pp. 371–404). Reston, VA: NCTM.

Hudicourt-Barnes, J. (2003). The use of argumentation in Haitian Creole science classrooms. *Harvard Educational Review, 73*(1), 73–93.

Lemke, J. (1990). *Talking science: Language, learning, and values.* Norwood, NJ: Ablex.

Mestre, J. (1988). The role of language comprehension in mathematics and problems solving. In R. Cocking & J. Mestre (Eds.), *Linguistic and cultural influences on learning mathematics* (pp. 259–293). Hillsdale, NJ: Lawrence Erlbaum.

Morgan, C. (2004). Word, definitions and concepts in discourses of mathematics, teaching and learning. *Language and Education, 18*, 1–15.

Moschkovich, J. N. (1999a). Supporting the participation of English language learners in mathematical discussions. *For the Learning of Mathematics, 19*(1), 11–19.

Moschkovich, J. N. (2000). Learning mathematics in two languages: Moving from obstacles to resources. In W. Secada (Ed.), *Changing the faces of mathematics: Perspectives on multiculturalism and gender equity* (Vol. 1, pp. 85–93). Reston, VA: NCTM.

Moschkovich, J. N. (2002). A situated and sociocultural perspective on bilingual mathematics learners. *Mathematical Thinking and Learning, 4*(2 & 3), 189–212.

Moschkovich, J. N. (2007a). Beyond words to mathematical content: Assessing English Learners in the mathematics classroom. In A. Schoenfeld (Ed.), *Assessing Mathematical Proficiency* (pp. 345–352). New York: Cambridge University Press.

Moschkovich, J. N. (2007b). Bilingual mathematics learners: How views of language, bilingual learners, and mathematical communication impact instruction. In N. Nasir & P. Cobb (Eds.), *Diversity, equity, and access to mathematical ideas* (pp. 89–104). New York: Teachers College Press.

Moschkovich, J. N. (2007c). Examining mathematical Discourse practices. *For the Learning of Mathematics, 27*(1), 24–30.

Moschkovich, J. N. (2007d). Using two languages while learning mathematics. *Educational Studies in Mathematics, 64*(2), 121–144.

Moschkovich, J. N. (2009). *Using two languages when learning mathematics: How can research help us understand mathematics learners who use two languages?* Research Brief and Clip, National Council of Teachers of Mathematics, available online at http://www.nctm.org/uploadedFiles/Research_News_and_Advocacy/Research/Clips_and_Briefs/Research_brief_12_Using_2.pdf.

Moschkovich, J. N., & Nelson-Barber, S. (2009). What mathematics teachers need to know about culture and language. In B. Greer, S. Mukhopadhyay, S. Nelson-Barber, & A. Powell (Eds.), *Culturally responsive mathematics education* (pp. 111–136). New York: Routledge, Taylor & Francis Group.

O'Connor, M. C. (1999). Language socialization in the mathematics classroom. Discourse practices and mathematical thinking. In M. Lampert & M. Blunk (Eds.), *Talking mathematics* (pp. 17–55). New York: Cambridge University Press.

O'Halloran, K. L. (1999). Towards a systemic functional analysis of multisemiotic mathematics texts. *Semiotica, 124*(1/2), 1–29.

O'Halloran, K. L. (2000). Classroom discourse in mathematics: A multisemiotic analysis. *Linguistics and Education, 10*(3), 359–388.

O'Halloran, K. L. (2005). *Mathematical discourse: Language, symbolism and visual images.* London: Continuum.

Planty, M., Hussar, W., Snyder, T., Provasnik, S., Kena, G., Dinkes, R., Kewal Ramani, A., & Kemp, J. (2008). *The condition of education 2008* (NCES 2008-031). Washington, DC: National Center for Education Statistics, Institute of Education Sciences, U.S. Department of Education.

Rosebery, A., Warren, B., & Conant, F. (1992). Appropriating scientific discourse: Findings from language minority classrooms. *The Journal of the Learning of Sciences, 2*(1), 61–94.

Rowland, T. (1999). *The pragmatics of mathematics education: Vagueness in mathematical discourse.* New York: Routledge, Taylor & Francis Group.

Schleppegrell, M. (2007). The linguistic challenges of mathematics teaching and learning: A research review. *Reading & Writing Quarterly, 23*, 139–159.

Secada, W. (1991). Degree of bilingualism and arithmetic problem solving in Hispanic first graders. *Elementary School Journal, 92*(2), 213–231.

Setati, M. (1998). Code-switching and mathematical meaning in a senior primary class of second language learners. *For the Learning of Mathematics, 18*(1), 34–40.

Setati, M., & Adler, J. (2001). Between languages and discourses: Code switching practices in primary classrooms in South Africa. *Educational Studies in Mathematics, 43*, 243–269.

Solano-Flores, G. (2006). Language, dialect, and register: Sociolinguistics and the estimation of measurement error in the testing of English-language learners. *Teachers College Record. 108*(11), 2354–2379.

Spanos, G., Rhodes, N. C., Dale, T. C., & Crandall, J. (1988). Linguistic features of mathematical problem solving: Insights and applications. In R. Cocking & J. Mestre (Eds.), *Linguistic and cultural influences on learning mathematics* (pp. 221–240). Hillsdale, NJ: Lawrence Erlbaum.

Stein, M. K., Smith, M. S., Henningsen, M. A., & Silver, E. A. (2000). *Implementing standards-based mathematics instruction: A casebook for professional development.* New York: Teachers College Press.

Tafoya, S. M. (2002). The linguistic landscape of California schools. *California Counts, 3*, 1–15.

Vygotsky, L. S. (1987). *The collected work of L. S. Vygotsky.* (R. W. Rieber, Ed., & N. Minick, Trans.). New York: Plenum Press.

Warren, B., Ballenger, C., Ogonowski, M., Rosebery, A., & Hudicourt-Barnes, J. (2001). Re-thinking diversity in learning science: The logic of everyday language. *Journal of Research in Science Teaching, 38*, 529–552.

Warren, B., Ogonowski, M., & Pothier, S. (2005). "Everyday" and "scientific": Rethinking dichotomies in modes of thinking in science learning. In R. Nemirovsky, A. Rosebery, J. Solomon, & B. Warren (Eds.), *Everyday matters in science and mathematics: Studies of complex classroom events* (pp. 119–148). Mahwah, NJ: Lawrence Erlbaum Associates.

Zentella, A. (1981). Tá Bien, you could answer me en cualquier idioma: Puerto Rican codeswitching in bilingual classrooms. In R. Durán (Ed.), *Latino language and communicative behavior* (pp. 109–131). Norwood, NJ: Ablex.

AFTERWORD

Beth Warren

An afterword is an occasion for both appreciation and reflection. I would like to do a bit of both here, the latter with a broad brush.

As a learning sciences researcher with a deep interest in language, particularly in relation to learning and teaching in the sciences, but with no claim to expertise in mathematics, I can testify to the educative power of the chapters in this volume. Individually and collectively, they chart a multilayered territory of theory, problems, arguments, examples, and questions for research on language in mathematics education, both broadly and specifically conceived. Notably, the volume as a whole locates matters of equity and justice at the center of the study of learning and development rather than at the periphery. This volume will, I believe, stand as a touchstone of the kinds of interdisciplinary research needed to generate a theoretically and pedagogically expansive understanding of the meanings, forms, and functions of language in mathematical learning and teaching in an increasingly multimodal, transcultural world.

As I read the chapters, I found myself thinking of Toni Morrison's 1993 Nobel Lecture (Morrison, 1999). It is an extraordinary meditation on what *doing language*, a main crosscutting theme of this book, means for our collective humanity. Among other possibilities, I think of the lecture as a meditation on pedagogy—specifically, a pedagogy rooted in deep, generative work in and with language to explore and create possible worlds.

Language and Mathematics Education, pages 171–174
Copyright © 2010 by Information Age Publishing
All rights of reproduction in any form reserved.

The lecture is framed around an age-old story, one found in the lore of many communities, of an encounter between a group of young people and an old woman, who is "blind but wise" and, in the version Morrison recounts, "the daughter of slaves, black, American," who "lives alone in a small house outside of town." One day she is visited by a group of young people. One of them asks the one question, which, because of her blindness, the woman seemingly cannot answer: "Is the bird I am holding living or dead?" After a long silence, the woman responds: "I don't know whether the bird you are holding is dead or alive, but what I do know is that it is in your hands. It is in your hands." (Morrison, 1999, pp. 9–11)

Morrison reads the bird-in-the-hand as language and the elder woman as a writer, who worries about the fate of language, whether it will live or die, a responsibility held by all language users and makers. Dead language, Morrison explains, is unyielding, unreceptive to new ideas, possible stories; it preserves hierarchies of inequality and privilege. Living language offers different, transformative possibilities, at once complicated and demanding:

> The vitality of language lies in its ability to limn the actual, imagined and possible lives of its speakers, readers and writers. . . . It arcs toward the place where meaning may lie . . . [i]ts force, its felicity is in its reach toward the ineffable. (Morrison, 1999, p. 20)

Readers may ask what this meditation on literary art has to do with language in relation to mathematics learning and teaching? Is not the ineffable—that which cannot be named—antithetical to mathematical discourse? My response, likely unsatisfactory, is this. As we move to investigate the kinds of deeply considered pedagogies argued for in this volume, we should take care to attend closely to the ways in which multifaceted work with the languages of mathematics can lead to *unforeseen* possibilities of mathematical meaning and purpose, to emergent spaces linking historically valued or established practice and creativity, precision and ambiguity, certainty and uncertainty, logic and intuition in ongoing learning activity (Gutiérrez, Baquedano-Lopez, & Tejeda, 1999; Lee, 2007; Nasir, Rosebery, Warren, & Lee, 2006; Warren, Rosebery, & Pothier, 2006). "Word-work," Morrison writes, "is sublime . . . because it is generative" (1999, p. 22). Word-work in mathematics, by analogy, ought to be as well.

In the last part of the lecture, Morrison has us imagine that the young people have nothing in their hands. She invites the reader to hear their question, not as a ruse, but as a plea to their elder for mutually engaged, open-hearted, and risk-taking boundary-crossing.

> We have no bird in our hands, living or dead. We have only you and our important question. Is the nothing in our hands something you could not bear to contemplate, to even guess? Don't you remember being young when

language was magic without meaning? When what you could say, could not mean? When the invisible was what imagination strove to see? When questions and demands for answers burned so brightly you trembled with fury at not knowing? (Morrison, 1999, p. 25)

Here, even as the young people chastise their elder for putting them off, they entreat her to risk sharing with them what she knows, how she has lived and struggled, even if her "reach exceeds her grasp," so that they may learn from her and her "particularized world" (Morrison, 1999, p. 27). They seek to ignite in the woman, across boundaries of experience, what they perceive as a lost intimacy with language as a way of being, feeling, and creating shared meaning.

Coincident with writing this afterword, a review by Jay Lemke (2009) of Anna Sfard's (2008) book, *Thinking as Communicating*, which figures prominently in this volume, crossed my desk. In his review, Lemke linked reasoning and feeling to learning, conceptualized within a communicational framework. Lemke's point is that communicative activity in mathematics, as in any discipline, is as much about representation and meta-representation as it is about "how we *feel* about what we're doing" (p. 283). One sense of feeling as it shapes learning, in this case professional learning, can be seen in the work of the biologist, Barbara McClintock. As is well known, she described her learning practice as "a feeling for the organism" (Keller, 1983, p. 198), a kind of intimate practice of knowing grounded in attentiveness, intuition, imagination, curiosity, and respect. Here is one example of McClintock describing her relationship to the corn plants she studied:

> No two plants are exactly alike. They're all different, and as a consequence, you have to know that difference. I start with the seedling, and I don't want to leave it. I don't feel I really know the story if I don't watch the plant all the way along. So I know every plant in the field. I know them intimately, and I find it a great pleasure to know them. (Quoted in Keller, 1983, p. 198)

McClintock imagines herself into the life of these plants. She knows them intimately. She finds pleasure in knowing them. She is alert to possibilities, to the unexplained. She knows, too, that these plants, all of the same kind, are also "all different."

What is the same from one perspective is also potentially different from another, understood not as a contrast of binaries but relationally. In this light, generative word-work in math, of the kinds envisioned in this volume, may be thought of as borderline engagements with manifold possible relations (Bakhtin, 1981; Bhabha, 1994; Nemerov, 1969). By emphasizing the relational character of word-work, in mathematics as in any discipline, I hope to convey the idea that all words, texts, images, symbols and actions evoke resonances, both individual and collective, as they "brush up against thousands of living dialogic threads" (Bakhtin, 1981, pp. 276–277). Living dialogic threads, coming as they do from many angles at once, including the culturally and historically saturated practices of academic disciplines, schooling, social communities, particular teachers, and particular students, make up the deep curriculum of pedagogies based in word-work, in mathematics no less than in science or literature or art. For pedagogies based in word-work, a crosscutting theme in this volume, this view of the fundamentally dialogic nature of language argues for exploration and experimentation that expands, rather than limits, the mathematical and other disciplinary discourses with which learners routinely engage.

REFERENCES

Bakhtin, M. (1981). *The dialogic imagination.* Austin, TX: University of Texas Press.

Bhabha, H. (1994). *The location of culture.* London: Routledge.

Gutiérrez, K., Baquedano-Lopez, P., & Tejeda, C. (1999). Rethinking diversity: Hybridity and hybrid language practices in the third space. *Mind, Culture, and Activity, 6*(4), 286–303.

Keller, E. F. (1983). *A feeling for the organism: The life and work of Barbara McClintock.* New York: W. H. Freeman and Company.

Lee, C. D. (2007). *Culture, literacy and learning: Taking bloom in the midst of the whirlwind.* New York: Teachers College Press.

Lemke, J. L. (2009). Learning to mean mathematically (Book Review). *Mind, Culture, and Activity, 16,* 281–284.

Morrison, T. (1999). *The Nobel lecture in literature, 1993.* New York: Alfred A. Knopf.

Nemerov, H. (1969, Autumn). On metaphor. *The Virginia Quarterly Review, 45,* 621–636.

Nasir, N. S., Rosebery, A. S., Warren, B., & Lee, C. D. (2006). Learning as a cultural process: Achieving equity through diversity. In R. K. Sawyer (Ed.), *The Cambridge handbook of the learning sciences* (pp. 489–504). New York: Cambridge University Press.

Sfard, A. (2008). *Thinking as communicating.* Cambridge, UK: Cambridge University Press.

Warren, B., Rosebery, A. S., & Pothier, S. (2006, April). *Working a metaphor: Borderline engagement with the unforeseen.* Paper presented at the Annual Meeting of the American Educational Research Association, San Francisco, CA.

CONTRIBUTORS

Jack Dieckmann

Dr. Jack Dieckmann is a Post Doctoral Scholar at the Stanford University School of Education and a Visiting Assistant Professor in the Research on Cognition and Mathematics Education unit at University of California, Berkeley. With funding from the Carnegie Corporation of New York and a Gerald J. Lieberman Fellowship, Dieckmann has recently conducted a two-year study that explores the relationship and concurrent validity between value-added statistical models of teaching effectiveness based on student achievement and observed patterns of use of semiotic mathematical resources in Algebra and Geometry classrooms. His previous research in teacher development investigated the range of guiding images that new mathematics teachers hold for equitable practice, as well as the tensions teachers face as they adopt critical perspectives about their practice and the institutions within which they work. In addition to his academic work, Dieckmann also serves as an advisor to the Migrant Education Program on policy matters at the state and national levels.

Kris D. Gutiérrez

Dr. Kris D. Gutiérrez is Professor of Literacy and Learning Sciences and holds the Inaugural Provost's Chair at the University of Colorado, Boulder. She is also Professor Emerita of Social Research Methodology in the Graduate School of Education & Information Studies at the University of California, Los Angeles where she also served as Director of the Education Studies Minor and Director of the Center for the Study of Urban Literacies. She is a Fellow of the American Educational Research Association, the Na-

Language and Mathematics Education, pages 175–180
Copyright © 2010 by Information Age Publishing
175

tional Conference on Research on Language and Literacy, and the Education and the Public Interest Center. Professor Gutiérrez's research has been published widely in premier academic journals such as *Review of Research in Education, Educational Researcher, Human Development, Mind, Culture and Activity, Reading Research Quarterly,* the *Harvard Educational Review, Journal of Literacy Research, Linguistics and Education, Discourse Processes, Research in the Teaching of English, Pedagogies,* and the *Journal of Teacher Education,* for example. She is a co-editor of the book, *Learning and Expanding with Activity Theory,* published by Cambridge University Press. Gutiérrez was elected to the National Academy of Education and nominated by President Obama to serve as a member of the National Board for the Institute of Education Sciences. She has received numerous awards, including the 2010 AERA Hispanic Research in Elementary, Secondary, or Postsecondary Education Award, the 2010 Inaugural Award for Innovations in Research on Diversity in Teacher Education, Division K (AERA), the 2007 AERA Distinguished Scholar Award, and the 2005 AERA Division C Sylvia Scribner Award for influencing the field of learning and instruction. She was a Fellow at the Center for Advanced Studies in the Behavioral Sciences in 2006–07 and the 2010 Osher Fellow at the Exploratorium Museum of Science. She serves on numerous policy-making and advisory boards. She served as a member of the U.S. Department of Education Reading First Advisory Committee and as a member of President Obama's Education Policy Transition Team. Professor Gutiérrez is the President for the American Educational Research Association (2010–2011), and President of the National Conference on Research on Language and Literacy (2009–2011). Gutiérrez has held Noted Scholar positions in Japan and Canada and is an invited speaker both nationally and internationally.

Judit N. Moschkovich

Dr. Judit N. Moschkovich is Professor of Mathematics Education in the Education Department at the University of California at Santa Cruz. Her research has used sociocultural approaches to mathematical thinking and learning to examine three topics: algebraic thinking; mathematical discourse practices; and bilingual mathematics learners, especially Latino/a students. Her publications have focused on student understanding of algebraic and graphical representations of functions and mathematical discourse practices in classrooms and everyday settings. She has conducted classroom research in secondary mathematics classrooms with Latino/a students, examined mathematical discussions among bilingual Latino/a students, and explored the relationship between language(s) and learning mathematics. Dr. Moschkovich was the co-editor, with M. Brenner, of JRME monograph Number 11, "*Everyday and academic mathematics: Implications for the classroom*" (2002). She has published articles in *The Journal of the Learning Sciences, Educational Studies in Mathematics,* and *For the Learning of Mathematics.* Recent book chapters include "Ecological perspectives on mathematical reasoning practices in Latino/a communities in the border-lands" (2010, in the volume edited by R. Kitchen and M. Civil, *Transnational and Borderland Studies in Mathematics Education,* Routledge, Taylor & Francis Group), "How language and graphs support conversation in a bilingual mathematics classroom" (2009, in the volume edited by R. Barwell, *Multilingualism in Mathematics Classrooms: Global Perspectives,* Multilingual Matters Press), and "Beyond words to mathematical content: Assessing English Learners in the mathematics classroom (2007, in the volume edited by A. Schoenfeld, *Assessing Mathematical Proficiency,* Cambridge University Press). She was the Principal Investigator of a National Science Foundation project (1998–2003) titled "*Mathematical discourse in bilingual settings: Learning mathematics in two languages*" and one of the PIs for a Center for Learning and Teaching funded by NSF (2004–2010), the Center for the Mathematics Education of Latinos/as (CEMELA). She has served on the Editorial Panel for the *Journal for Research in Mathematics Education,* the Review Board for *The Journal of the Learning Sciences,* and as the Chair for the *Research in Mathematics Education SIG* (Special Interest Group) in AERA (American Educational Research Association). She currently serves as a member of the International Program Committee of the International Council for Mathematics Instruction (ICMI) Study #21: *Mathematics education and language diversity.*

David Pimm

Dr. David Pimm is currently Professor of Mathematics education in the Dept of Secondary Education at the University of Alberta, Canada, where he has worked since 2000. For much of his academic career (1983–1997), he worked in the Department of Mathematics at the Open University in the UK, before spending two years in the Department of Teacher Education at Michigan State University. His primary interests involve examining and exploring the many aspects of language use in mathematics classrooms (both spoken and written) at all levels, in addition to aspects of the history and philosophy of mathematics, both as areas of study in themselves and in relation to pedagogy. The linguistic basis for mathematical activity is a further preoccupation. In his thirty-plus year career to date, Pimm has published widely, including three authored or co-authored books and six edited or co-edited volumes, the most recent being *Mathematics and the Aesthetic: New Approaches to an Ancient Affinity* (Springer, 2006). He was also the editor of the international journal *For the Learning of Mathematics* for six years. His current work involves an exploration of the poetic in relation to mathematics, focusing in particular, but not exclusively, on aspects of metaphor and metonymy, not least as a means to refocus on the role of the specific in mathematical work.

Mary J. Schleppegrell

Dr. Mary Schleppegrell is a Professor of Education at the University of Michigan School of Education. Her research explores the relationship between language and learning with a focus on students for whom English is a second language. A linguist, she uses the theoretical framework of systemic functional linguistics and its functional grammar to link meaning and language structure in ways that illuminate issues in education. She has written on the challenges of science, history, language arts, and mathematics, drawing implications for teacher education and K–12 classroom practice. Her work on the language challenges of mathematics has been published in *Reading and Writing Quarterly* (2007) and her collaboration with Beth Herbel-Eisenmann on discourse analysis of mathematics classroom language is reported in NCTM's *Mathematics for every student: Responding to diversity, Grades 6–8* (2008). With Zhihui Fang, she has written *Reading in Secondary Content Areas: A Language-Based Pedagogy*, (2008, University of Michigan Press), which includes a chapter on language in mathematics. She has recently written a chapter titled "Linguistic tools for exploring issues of equity in mathematics"; to be published in a forthcoming volume edited by Jeff Choppin, Beth Herbel-Eisenmann and David Wagner. Her other publications include *The Language of Schooling* (Erlbaum, 2004); and with Cecilia Colombi, *Developing Advanced Literacy in First and Second Languages: Meaning with Power* (Erlbaum, 2002). Her work has appeared in *Linguistics and Education*, the *Journal of Literacy Research, TESOL Quarterly*, and other journals and edited volumes.

Tesha Sengupta-Irving

Dr. Tesha Sengupta-Irving is the Assistant Director of Research at the UCLA Lab School and is a Lecturer and Postdoctoral Fellow in the Graduate School of Education & Information Studies at the University of California, Los Angeles. Her research draws on sociocultural and feminist research frameworks to examine the social organization of learning in mathematics classrooms. She focuses on group work and other forms of peer collaboration as unique opportunities to learn mathematics, and as windows into the ways gender, language, ethnicity, and academic ability shape student interactions and accomplishments. Her prior study, sponsored by the Spencer Foundation and a Morgridge Family Fellowship, examined how a class of academically "low track," predominantly female, ethnic minority students responded to group work as a pedagogical reform aiming to improve their affective and academic orientation to Algebra.

Guillermo Solano-Flores

Dr. Guillermo Solano-Flores specializes in educational measurement, assessment development, and the linguistic and cultural issues that are relevant to both the testing of linguistic minorities and international test comparisons. He is Associate Professor of Bilingual Education and English as a Second Language at the School of Education of the University of Colorado, Boulder. A psychometrician by formal training, his work focuses on the development of alternative and multidisciplinary approaches that address linguistic and cultural diversity in testing. He has conducted research on the development, translation, localization, and review of science and mathematics tests; the design of software for computer-assisted scoring; and the development of assessments for the professional certification of science teachers. He has been principal investigator in several National Science Foundation-funded projects that have examined the intersection of psychometrics and linguistics in testing. He is the author of the theory of test translation error, which addresses testing across cultures and languages. Also, he has investigated the use of generalizability theory—a psychometric theory of measurement error—in the testing of English language learners. He has advised Latin American countries on the development of national assessment systems and contributed to the development of the National Assessment of Educational Progress 2009 Science Framework with advice on strategies for the testing of linguistic minorities. Current research projects investigate the measurement of mathematics academic language load in tests and the design and use of illustrations as a form of testing accommodations for English language learners with an approach that uses cognitive science, semiotics, and sociolinguistics in combination. He is a member of the research team of an international study that investigates the feasibility of adapting and translating performance tasks into multiple languages.

Beth Warren

Dr. Beth Warren is Principal Scientist and co-Director of the Chèche Konnen Center at TERC, a not-for-profit education research and development organization in Cambridge, MA. Her research integrates four strands of inquiry: a) documentation of the wide-ranging, intellectually powerful sense-making repertoires of students from historically non-dominant communities, repertoires that in schools and the larger society are usually not recognized as academically meaningful, b) analysis of generative intersections between these sense-making repertoires and those used routinely in the everyday work of academic disciplines, such as the sciences, humanities and arts, c) in collaboration with classroom teachers, exploration and development of classroom practices that build on heterogeneity as a first principle of design for expansive learning, and d) in collaboration with classroom teachers, exploration and development of an approach to professional development, called learning-in-practice, which integrates investigations of subject matter, learning, pedagogy, and historically structured inequalities on the same plane of professional inquiry. This research has been funded by the National Science Foundation, the Spencer Foundation, the U.S. Department of Education, and the Ford Foundation, and has been published widely in books and journals. Dr. Warren currently serves on the Editorial Board of *Cognition and Instruction*.